Jürgen Kranz

Comparative Epidemiology of Plant Diseases

Springer

Berlin
Heidelberg
New York
Hong Kong
London
Milan
Paris
Tokyo

Jürgen Kranz

Comparative Epidemiology of Plant Diseases

With 64 Figures

Springer

Professor Dr. Jürgen Kranz
Rehschneise 75
35394 Giessen
Germany

ISBN 3-540-43688-X Springer-Verlag Berlin Heidelberg New York

Library of Congress Cataloging-in-Publication Data

Kranz, Jürgen.
   Comparativ epidemiology of plant diseases / Jürgen Kranz.
      p. cm.
   Includes bibliographical references.
   ISBN 354043688X (alk. paper)
     1. Plant diseases--Epidemiology. I. Title.

SB731 K69 2002
632'.3--dc21   2002026691

Springer-Verlag Berlin Heidelberg New York
a company of BertelsmannSpringer Science+Business Media GmbH

http://www.springer.de

© Springer-Verlag Berlin Heidelberg 2003
Printed in Germany

Cover design: *design & production* GmbH, Heidelberg
Cover illustration: Cover photograph by J. Kranz
Typesetting: Camera-ready by Steingraeber Satztechnik

SPIN 10864472   31/3130 YK – 5 4 3 2 1 0 – Printed on acid free paper

Preface

Comparison is an important cognitive research tool in science. Results in practically every study are compared with various hypotheses, other results or practices. These comparative "within-studies" advance knowledge, solve queries or problems and improve practical applications. Comparative research, in contrast, examines "across-studies" to evaluate similarities and differences between units, e.g. across taxa of organisms. Ever since Charles Darwin's comparisons "across", e.g. across species, they have proven to be an effective approach in research in ecology, phylogenetics, palaeontology, systematics, ethology (animal behaviour) and in plant disease epidemiology. Comparisons "across" are made by observations, experiments, or by posterior analysis with data of experiments or observations. Properly applied, comparative research with appropriate methodology will not only increase or modify knowledge, but will also identify essentials, generalisations, and restrictions to consolidate the foundation of a science. It thus establishes hypotheses, theories, principles and even laws, and comparison can be used for their critical analysis.

This text is devoted to comparative epidemiology of plant diseases, which has developed in its own way along with human and animal epidemiology. Though not written as a review of its status, it rather attempts to offer an introduction as to how to conduct comparative research, for whatever reason. Comparative epidemiology is essentially a study across pathosystems, epidemics and their relevant aspects. It is a powerful tool with great potential to advance plant disease epidemiology as a science and improve crop protection. The text addresses graduate students in plant pathology, their teachers, epidemiologists, professionals in plant pathology and crop protection and to readers of ecology and similar branches of applied biology.

With pleasure I acknowledge critical and helpful comments during the preparation of the manuscript from Armando Bergamin Filho, Richard D. Berger and Bernhard Hau, who read the drafts of the manuscript and made valuable contributions. I also thank Wolfgang Köhler, Gabriel Schachtel and Josef Rotem for useful comments and suggestions. Thanks also go to Ms. Karin Schlosser who prepared the figures.

Jürgen Kranz
Giessen, September 2001

Contents

1 Introduction

Why comparative epidemiology of plant diseases? Zadoks and Schein (1980) rightly stated that results from individual experiments and observations tend to catch and encapsulate scientists who then might generalize beyond their results; this often leads to far-reaching, but premature hypotheses on epidemiological aspects. More consolidated hypotheses, however, can be obtained when diseases and their epidemics are compared with others. Comparative epidemiology (CE) is called upon to test such hypotheses by falsification in appropriate experiments or posterior analyses. Also, hypotheses that arise from observation, professional experience and reviews may be tested in such a way.

Falsification, sensu Popper (1973), searches for errors to disprove and negate existing concepts, e.g. theories. The assumption "all swans are white" is falsified by a single black one. Kuhn (1978), however, warns of any naive use of Popper's approach (e.g. the swan example) because a theory must not necessarily be valid for all possible applications. The theory simply may not fit specific conditions. It may suffice to redefine a theory or make it more specific. If, however, comparative research reveals discrepancies between theory and new facts which can no longer be reconciled, then a change of paradigm will occur and previously held views and theories will be abolished. These changes may be slow and gradual, or abrupt, becoming something entirely new. Comparative research implies reduction, usually favoured by statistical methods. Popper (cited by Lorenz 1978) regards this as a scientific success, which, even if it fails, still leaves behind challenging questions as intellectual properties of research.

Comparative epidemiology essentially does across-studies of pathosystems (p. 7) and the temporal as well as spatial aspects of their epidemics and structural elements. The same pathosystem may also be compared across different characteristic environmental conditions (e.g. at pathotopes sensu Putter, pers. comm.; see Sect 6.4) and among distinct methods of system control or design (see Chap. 7). The latter comparisons are often made in the course of validating experimental results, though not necessarily seen in the context of comparative epidemiology.

A pathosystem results from interactions in the disease square (Zadoks and Schein 1979, based on Van der Plank's fungicide square of 1963) comprising the pathogen (or causative agents such as deficiencies, etc.), the host, the environment and human interference. A pathogen may be assisted by other organisms to cause disease, for instance, vectors, predisposing fungi or helper viruses, which become part of the pathosystem. Weather factors and human actions, e.g. agricultural practices, can substantially influence epidemics. Therefore, their effects may also be studied across climatic conditions or agricultural practices, even within the same pathogen-host combination, e.g. across pathotypes, different sites or agri-

cultural practices. Data obtained from the validation of an experimental comparison, usually done under varying climatic and agricultural conditions, may be used
for posterior comparisons.

With comparative epidemiology, differences or similarities across studies of
pathosystems or epidemics are examined with specifically designed experiments
(Sect. 3.2.1) or posterior analyses (Sect. 3.2.2). By means of these two approaches,
generalizations are attempted from the great diversity of epidemics of hundreds of
diseases that occur under a variety of environmental conditions and agricultural
practices. New concepts, hypotheses, theories, principles and laws may be established. Existing ones may be tested to determine if they are still valid against new
facts and developments. All this is instrumental in placing epidemiological aspects
into a meaningful context.

The comparison of epidemics is a method for both analytical and synoptic research. Comparative epidemiology compares underlying principles of epidemics
across studies of pathosystems, their epidemics and factors affecting them. Comparative epidemiology thus plays a "...unifying and crystallizing role...," (Butt and
Royle 1980) as it distils commonalities or differences in behaviour and structures
which help to explain why they exist. From the apparently unlimited diversity of
epidemics, a convenient number of basic types of epidemics may emerge eventually to which new epidemics could then be assigned (Kranz 1978, 1988b). This
process would help to consolidate epidemiology and to pave the way for research
to reach conclusions more speedily and with fewer detours. The plant pathologists
then would see just the forest, rather than a multitude of trees.

Comparative studies differ in branches of science in their philosophy, objectives and procedures. They have, when appropriately adjusted, some general features. For instance, in phylogenetics and systematics, comparative studies can be
stripped down to the following five elements (Gittleman and Luh 1992): (1) The
main hypothesis for the objective with or without causal explanations, e.g. ecological or evolutionary factors affecting phenotype, or co-variation between two
criteria (traits); (2) the range of variation in the criteria or traits (in terms of standard deviation); (3) the presence, location and form of any correlation relating to
the objective of the study; (4) the range of variation once the trait(s) under study
have been transformed through some comparative statistical procedure, e.g. to re-
move correlations; and (5) the knowledge of the difference in rate of change
(among the criteria used) that will impinge on the divergence of criteria. Properly
adopted, these points are valid also for comparative epidemiology.

As a general development, however, comparative studies now may be "un-
avoidably statistical" (Gittleman and Luh 1992). This is because of the accumulation of basic data on many traits in many areas of science (e.g. behavioural science
and ecology) and the availability of computers with ample capacity and suitable
software. Also, the knowledge basis and framework for comparative research have
become more solid than ever. All this leads to a better understanding of the intri-
cacies and the background of the object under study. Statistical tests, where applicable, will reveal more objectively the importance of variants under comparison,
i.e. those variants that are significantly similar or different from others, in particular, from standards or hitherto accepted concepts. Statistical methods also have
reductional properties. A brief guideline on the use of statistical methods in com-

parative epidemiology can be found, together with other methods for comparison, in Section 3.3.

As biologists are usually interested in various aspects of living systems per se, their reductionism and abstraction may not go so far as, for instance, in physics. Comparison in biology may require the perception of entirety (Gestalt) which is more than the sum of its parts. For Koehler and v. Bertalanffy (cited by Lorenz 1978) "Gestalt" is the "...harmonic and – effective in both directions – interlinked causal chains, the harmonic interaction of which causes the entirety ". This sounds rather similar to the definition of a system (see p. 5) by Watt (1966). As a research guideline it helped ethologists, who were familiar with the range of possible patterns, to discover certain inherent patterns of animal motion in related animal taxa. Such a comprehensive perception can capture relations and configurations which, together with rational thinking and detailed studies, will help to discover unexpected principles by comparative research. For comparisons of behaviour and phylogeny, Wenzel (1992) likes to understand: "The very breath of life itself and the living world in all its richness demonstrates that the "whole" can be much more than the sum of its parts". He continues, "Although we must have a certain combination of genes to permit us to speak, there is likely no gene for speech." In epidemiology, the graphs of the temporal and spatial dynamics of epidemics (i.e. disease progress curve and gradient) may be taken as entireties. They will be dealt with in Chapters 5 and 6 as the result of many factors and reactions that interact. Robinson (1976) had already referred to the term "Gestalt" in the context of pathosystems. With the system and the entire context in mind, results from rational comparisons, even on detailed aspects, can then be successful as long as they convey a grasp of the extent of phenomena, the problems involved, and a feeling for the diversity of possible views and interactions.

In plant pathology, Gäumann (1951) and Van der Plank (1963) first used comparison of epidemics and relevant factors to arrive at generalities about epidemics. In his *Principles of plant pathology* Yarwood (1973) lists 17 of his principles as related to epidemiology. For instance, he considered the poor relation which exists between the biotic potential of the pathogen and the resulting severity of the disease it causes, as a principle. He also gives this rank to the always differential response of host and pathogen to environment and to the number of diseases that increase with the increasing production of a crop. An experimental comparison of epidemics for more generally applicable information was described by Kranz (1968a–c, 1974a, 1978). A range of possible applications of comparative epidemiology is presented in the volume edited by Palti and Kranz (1980). In this volume, Zadoks and Schein (1980) distinguish the individual, the population and the community level in epidemiology and choose a comparison of processes, objectives and tools. Aust et al. (1980) compared the ability of factors in the disease triangle that act on the system "epidemic" to compensate for each other. Thresh (1980) reviewed the effect of factors to compare virus diseases and Jones (1980) did so for nematodes. Rotem and Palti (1980) compared the efficacy of cultural versus chemical disease control and Putter (1980) discussed factors relevant to the management of outbreaks of endemic diseases under tropical subsistence farming.

After a brief introduction to plant disease epidemiology and a more detailed definition of comparative epidemiology with its aims and scope in Chapter 2,

Chapter 3 deals with its methodology. Posterior analyses (Sect. 3.2.2) should make use of the extensive amount of information that has accumulated from many experiments, which would otherwise just pile up or be buried in "data cemeteries" and collections. This unused and unrelated information is a treasure with which objectives of comparative epidemiology can be achieved. Chapter 4 is devoted to the systems level of host and pathogens and, thus, to the elements as criteria for comparisons within the monocycle of epidemics. Chapter 5 deals with the temporal and Chapter 6 with the spatial aspects of epidemics. Finally, Chapter 7 describes how the effects of epidemics on crops can be compared. This obviously relates to disease management through systems control and system design. Comparative epidemiology then aims for a more rational use of resources to identify real research needs worth studying in epidemiology.

Examples of across-studies will be presented to demonstrate feasible applications of comparative epidemiology. Also, a few published "within-study" comparisons are cited to exemplify the range of possible comparisons, useful criteria and methods that might be adopted for comparative epidemiology. The emphasis of this text will be on criteria, methods and procedures. No exhaustive account of any across-study is given, but ample references to suitable publications are provided which can be used as guidelines for intended projects.

2 Plant Disease Epidemiology and the Scope of Across-Comparison

In the epidemiology of plant diseases, the dynamics of populations of pathogens in populations of hosts are studied along with the resulting disease under the influence of environmental factors and human interference. Epidemiology is the ecological branch of plant pathology and it is dealt with exhaustively in a number of textbooks: Zadoks and Schein (1979), Campbell and Madden (1990), Rapilly (1991), Nagarajan and Muralidharan (1995), Bergamin Filho and Amorim (1996), Kranz (1996). In all of these volumes, there are chapters on comparative epidemiology. The epidemiology of plant virus diseases is covered comprehensively in the volume edited by McLean et al. (1986) and papers by Thresh (1974a, 1976, 1983). For particular aspects of soilborne and root diseases, the volume by Campbell and Benson (1994a) should be consulted. The proceedings of a symposium edited by Palti and Kranz (1980) are devoted entirely to comparative epidemiology.

Epidemiology is obviously concerned with epidemics, that is any increase or decrease in disease intensity y in the range $0 < y \leq 1$ (or 100%) in time and space (Kranz 1974b). This comprises Gäumann's (1951) classical definition of epidemics adopted from human medicine, i.e. a steep temporal or spatial increase in disease followed by a decline within a limited period of time. Within the context of crop protection, epidemiology is the research interface between laboratory research and actual disease management in the field. Comparative epidemiology can provide information, amongst others, for the design of integrated pest management (IPM) schemes, from the behaviour of epidemics and the reactions of their components to weather factors and control measures. Finally, comparative epidemiology provides generally valid information for teaching and textbooks. For this more communicational application, an unequivocal language and terminology is particularly required (Sect. 3.1).

Through systems analysis, comparative epidemiology can be used to develop tactics and strategies for a more efficient, economic and sustainable management of disease. Plant diseases are open, coupled dynamic systems regulated by external factors (Kranz 1974b; Robinson 1976; Kranz and Hau 1980). The systems are interlocking processes determined by many reciprocal cause-and-effect relationships that characterise biological systems (Watt 1966), as is shown schematically in a simple relational diagram (Fig. 2.1). Stimuli affect the structural elements of disease progress curves (see also Fig. 5.2) and disease gradients directly, or through other (often preceding) elements, either positively or negatively. Depending on the type of stimuli and the weight they have on the various elements of the structure, different behaviour outputs are produced by the system. Flow charts of simulators (e.g. Waggoner and Horsfall 1969 for early blight of tomato), though reduced and simplified, present the internal relationships of epidemics affected by external factors. For infectious diseases, which make up the majority of plant diseases, only one incontestable cause-and-effect relationship exists in their epidemics: there is no disease without a pathogen and a susceptible host. Thereafter, each event in the disease cycle incites a following one: no infection will happen without inoculum and no inoculum without previous infection etc. The dynamics of such an internal process in systems is operated by features inherent in the populations of the host plants and the pathogens, like random processes, recurrence, limits, thresholds and discontinuities.

In the epidemiological hierarchy of pathosystems, pathogen and host are at the lowest systems level. From the interactions in the disease square of pathogen-host plant-environment-human interference, diseases result. At the next level, epidemics of a given disease in a crop are operated by spatio-temporal dynamics. The systems levels agro-ecosystem, agricultural area and levels above widen the geographical scope of epidemics and, as such, these systems are also relevant to comparative epidemiology. In agro-ecosystems, several diseases and crops occur under practically the same soil conditions, microclimate and agricultural practices, e.g. a field or farm. Agricultural areas comprise various agro-ecosystems under a similar mesoclimate and comparable agricultural conditions and practices. Any geographical delimitation may apply to the latter systems levels and any higher ones, such as global space. To work with systems, they first have to be bounded. This is essential in a systems approach and it implies a clear definition of what the

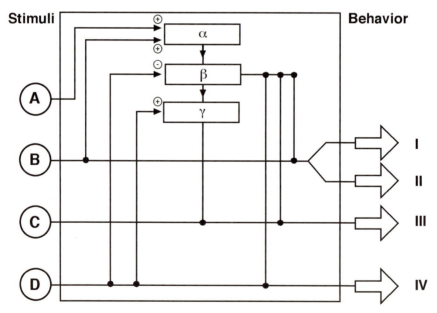

Fig. 2.1. A schematic of internal interactions in a system with inputs (stimuli) *A* to *D* and outputs (behavior) *I* to *IV*, affecting elements α, β and γ, their structure (indicated by *lines* and *dots*), and, differently, their effects (+, −). (Adapted from Hinde 1973; Kranz 1988b)

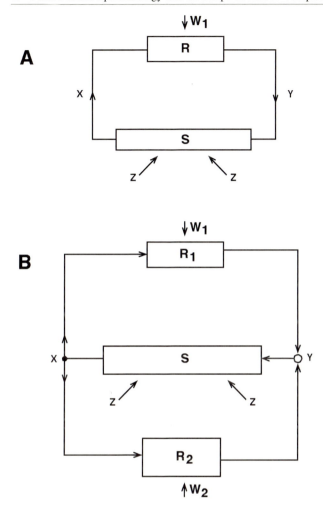

Fig. 2.2. Cybernetic relationships in **A** natural and **B** agricultural ecosystems from the point-of-view of pest control. W_1 is the natural tolerance level of disease (equilibrium) determined by factors acting in the natural ecosystem, W_2 is the desired disease intensity (e.g. economic threshold level) of the agro-ecosystem achieved by directed human interference, R represents the controlling natural factors, R_1 control by agricultural practices including chemical control, R_2 control by natural factors, y is the correcting variable (inoculum, new lesions), x is the controlled variable (disease intensity), S is the controlled system (either increase or decrease in disease intensity), z is the variable disturbance (e.g. unfavourable weather). (Aust et al. 1985)

system is going to be and how a system is delimited. Comparative epidemiology can be done either at the same systems level or between various levels.

A pathosystem, according to Robinson (1976), is "any subsystem of an ecosystem which involves parasitism". It is a host-pathogen combination with the ensuing interactions. Though "pathosystem" may be used as a synonym for disease, it will be applied here as a term for a pathogen-host-environment interaction, while "disease" refers to the population of symptoms manifested. Disease assessment is based on their symptoms, or lesions, on a host plant. According to Robinson's definition, powdery mildew (*Erysiphe graminis*) on adjacent fields of barley and wheat are two pathosystems (in this case as *formae speciales*). The three *Puccinia* spp. on wheat are considered as three separate pathosystems. Also, all different diseases in a crop are pathosystems of their own. Finally, comparisons on aspects and effects of factors (environmental or agricultural practice) across pathosystems are also important topics of comparative epidemiology (e.g. Chap. 4).

A pathosystem occurs on wild host plants and on crops. For pathosystems in naturally growing plant populations or communities, Van der Plank's disease triangle (host-pathogen-environment) applies, whereas in crops it is the disease square (Van der Plank 1963) which includes human interference (Chap. 7). Although diseases occur in both ecosystems, their basic structures differ for plant pathogens which leads to different feedbacks (Fig. 2.2). Epidemics in populations of natural host plants and crops have different cybernetic relationships. Disease intensities in populations of wild plants with the carrying capacity K (Sect. 4.1.3) of the individual host plants A entirely result from pathogenicity, host plant resistance and weather. Interventions by means of control measures S are lacking as these interventions may be applied in crops B either directly or by means of cultural practices in addition to environmental factors affecting W_2.

The scope of comparative epidemiology encompasses all system levels relevant to plant disease epidemiology at the micro-, meso- and macro-strata of epidemiological research (Kranz 1996). The micro-strata refers to the dynamics of intra-specific selection processes in pathogen and host populations (e.g. resistance-virulence interactions) prompting evolution in pathogen and host populations, which may alter epidemics of a pathosystem as dealt with in detail by Robinson (1976). Most epidemiological studies are at the meso-strata, the level of epidemics in plant populations and communities with their disease progress curves and gradients. The macro-strata of epidemiology corresponds to what is known as geophytopathology (Weltzien 1972), e.g. for comparisons of regional or global disease proneness and vulnerability (sensu Robinson 1976). These are factors which favour pandemics of diseases and the presence or absence of a disease in a region of the world.

Comparisons between different systems levels require related parameters and appropriate experimental approaches, e.g. when comparing the "sporulation" at the disease level to "spore dispersal" at the epidemic level. Meaningful comparisons are more easily possible at points of the different systems levels under comparison where the relevant criteria are shared among most of the levels.

3 On the Methodology
of Comparative Epidemiology

Successful comparative epidemiology requires a good knowledge and under-standing of etiology, the ecology of host and pathogen and the epidemics of the pathosystems under study. Comparative studies on hypotheses begin with a clearly formulated objective and the choice of essential and meaningful criteria. The methods to be used are the ones accepted and appropriate in epidemiological re-search as described, for instance, in Kranz and Rotem (1988) and Francl and Ne-her (1997) and in more general terms by Zadoks (1978). The importance of a proper terminology (Sect. 3.1) for comparative epidemiology was stressed by Butt and Royle (1980).

Comparisons based on data from sets of specifically designed experiments across pathosystems would be ideal (Sect. 3.2.1). Such comparisons, however, are rare and often difficult to make. Instead, published evidence, reports and data col-lected otherwise should be used for posterior analyses across studies to further epidemiology by comparative research (Sect. 3.2.2). Results derived from experi-ments and posterior analysis have to be validated by experiment, simulation or sometimes by experience only. To examine underlying hypotheses or theories, suitable statistical methods (Sect. 3.3.2) are available for hard, mild, and soft tests, sensu Zadoks (1978), and for falsification. Whenever possible and justified, re-searchers in comparative epidemiology should adopt a systems approach as their investigating procedure (Kranz and Hau 1980; Seem and Haith 1986).

3.1 The Importance of Proper Terminology

Vagueness, remoteness and ambiguity of terms and definitions limit communica-tion. Therefore, proper terminology should be employed to make conclusions of comparative epidemiology unequivocal. Butt and Royle (1980) also stressed the need for clear definitions of descriptive and measurement terms, parameters and notions in epidemiology as a quantitative science. A definition bounds and char-acterises a concept and allows it to be related to others. Explicit concepts among the definitions more easily ensure a broad acceptance, whereas more abstract con-cepts tend to be "ambiguous and vague" (Butt and Royle 1980). Whenever possi-ble, the adopted terms also should convey the sense of a phenomenon, its diversity and interactions (p. 3).

A *concept* is either an idea, a lasting mental image or a general term which em-braces all that is associated with, or suggested by, this term as a label of a concept, e.g. in the first column in Table 3.1. Concepts have various degrees of abstraction. Consider the following sequence: propagule, biflagellate zoospores, zoospores, spores, inoculum etc. A *hypothesis* is a supposition of facts or phenomena, a basis for reasoning and for the planning of research, although the truth of it still has to be proven (Butt and Royle 1980). From the statistical point of view, it may either be right or wrong, when tested for the null hypothesis. Hypotheses are usually starting points for new investigations, and then can be precursors or drafts of theories. A *theory* (like hypotheses) may just be a fanciful speculation without regard to known facts (Butt and Royle 1980). Established and accepted theories are rather a set of propositions that explain and illustrate the state of workable knowledge attained from facts or processes. In science, theories are a system of logically related and interdependent hypotheses. Theories in biology, however, are not as strictly formulated and postulated as theories would be in mathematics or physics. Among the various possible theories, the one theory that explains the real phenomenon better than any other will be adopted, even though often after considerable dispute. New facts may improve, restrict or even invalidate an accepted theory and lead to a new paradigm. Theories may encompass *principles* as fundamental truths, rules or basic facts, e.g. any infectious pathogen starts colonisation of its future host by infection. In physics, for instance, mass, power, energy and space are well-defined principles. Gäumann (1951) and Van der Plank (1963) established principles in plant disease epidemiology (Sect. 3.3.1). Some principles of plant pathology were listed by Yarwood (1973; p. 3). An established self-evident principle becomes *axiomatic*, e.g. there will be no rust disease without a pathogen from the Uredinales, or unilateral disease progress curves can be described by means of growth functions. A *law* in natural science, according to the Oxford Dictionary, is the correct statement of an invariable sequence between specified conditions and specified phenomena. The classical examples are Galilei's law of gravity and Newton's axioms. In epidemiology the infection chain for infectious diseases is a law, amongst others. Butt and Royle (1980) distinguished empirical laws that state consistencies between phenomena and general laws that summarise the theoretical "orderliness" of nature. As examples from epidemiology, they quote Van der Plank (1975) with the "law of origin" and the difference equation of epidemic growth, respectively. The notion *model* refers to an abstraction or a simplified approximation to the real world, or parts of it, as needed for a defined objective or purpose. This implies that a model rarely is complete or final (Kranz 1990a). It is often used as a somewhat loose term, even for mathematical functions, like the apparent infection rate or a linear regression functions. A model can have different forms (verbal, graphic, mathematical etc.). Their status is a conceptual model at the beginning of the investigation which, finally, may develop into a theory, principle or even a law. Models may be tools for descriptive purposes (e.g. the disease square) or operational programs (e.g. simulators, forecasters). Within the systems approach, models may be both output and research tool at each stage of their development. In comparative epidemiology, models can be useful tools to test hypotheses (Sect. 3.3.2). Organisms under controlled conditions (e.g. mini-epidemics in growth chambers, wind channels) are also used as models to study certain aspects or a hypothesis.

Table 3.1. Examples of descriptive terms (attributes) grouped according to levels of the pathosystem.[a] (Butt and Royle 1980)

Common names of entities[b]	Qualitative terms denoting properties	Behavioural terms denoting processes
A		
Propagule	Survivability	Dispersal
Saprophyte	Longevity	Dissemination
Spore	Persistence (virus)	Transport
Vector		Deposition
B		
Inoculum	Sporulation	Infection
Disease	Survivability	Incubation
Disease cycle	Viability	Sporulation
Monocycle	Infectiousness	Latency
C		
Infection chain	Epidemic	Disease progress
Polycycle	Endemic	Disease spread
Primary inoculum	Exodemic	Overwintering
Pathosystem	Polyetic	Removal

[a]Terms selected from Table 1 in Butt and Royle and slightly changed. The terms of quality, listed as examples here, reflect the hazards of dissemination from a source. Note: terms in the same line of Tables 3.1 and 3.2 are not to be related to each other.
[b]A, non-parasitic; B, parasitic; C, epidemic levels.

Table 3.2. Some measurement terms used at the parasitic level of the pathosystem with A as inoculum, B as temporal terms and C as inoculum production terms. (Butt and Royle 1980)

Class I	Class II	Class III
Terms precise and explicit	$\rightarrow$	Terms abstract, conceptual
At the non-parasitic level of the pathosystem		
Airborne spore conc. $[N \times L^{-3}]$	Deposition gradient	
Terminal velocity v $[L \times T^{-1}]$		
Trapping efficiency [1]		
At the parasitic level of the pathosystem		
A Infection efficiency [1]	Inoculum density	Effective inoculum dose
B Mills period [T]	Latent period, infectious period	Infection period
C Sporulation density $[N \times L^{-2}]$		Sporulation intensity
At the epidemic level of the pathosystems		
Infection rate r $[T^{-1}]$	Disease gradient	Disease severity
Sanitation ratio [1]	Dispersal gradient	Disease intensity
Disease gradient $[N \times L^{-1}]$	Disease intensity	Disease incidence

From Tables 2, 3 and 4. (Butt and Royle 1980). See also footnote to Table 3.1.

Criteria for comparative epidemiology are either descriptive or measurement terms (Tables 3.1 and 3.2), preferably with biological significance. They do not differ essentially between usage within-studies and across-studies for comparative epidemiology. However, each criteria, as shown in Tables 5.1 and 5.2 when employed in comparative epidemiology, may have a particular bearing or weight in across-studies of different pathosystems. Criteria adopted for the comparison of epidemics and their structures can be homologous or analogous. Homologous components are those that are congruent with another entity. They are phylogenetically alike, e.g. infection by means of conidia of haploid pathogens, or sporidia in the Ustilaginales. Analogues are structural components which are similar only in their functional mechanisms, like sclerotia or rhizomorphs as infective structures (Kranz 1978). Homology characterises similarity more strongly than analogy. In many areas of biology, research on homologues has led to useful conclusions, particularly if criteria have a distinct phylogenetic background. Usually, criteria of structures (e.g. morphological, anatomic or taxonomic attributes and parameters) tend to be more stable across a pathogen taxa etc. These criteria usually are more easily compared than those for the "more plastic" behavioural aspects (Wenzel 1992).

Descriptive terms as criteria are either technical terms of structures, symbols or verbal expressions such as common names and terms of quality or of behaviour (Table 3.1). These terms are the *attributes* (or traits) of characteristic qualities of objects under comparison which may be derived directly from perception (e.g. the "spore"). However, more abstract qualitative concepts, like dispersal, require more complex definitions. Attributes may have varying *features* as a qualitative notion of distinctive characteristic parts, e.g. monocyclic or polycyclic for "epidemic" (Table 3.1). *Aspect* is used as a loose term for a specific or passing appearance of an attribute.

From the descriptive criteria, of course, measurement terms can be derived with their parameters, e.g. spore = spore size in microns, latency = latent period p in days. Some examples adopted from Butt and Royle (1980) are listed in Tables 3.1 and 3.2. At the non-parasitic level A in these tables, the pathogen is not yet in contact with the host; these terms mainly relate to dissemination and survival. Active interaction at parasitic level B is taking place between pathogen and host. The term "infection cycle", for instance, leads to the concept of generation in pathosystems. The truly epidemic level C draws upon events at the lower levels A and B. The terms for entities, properties and processes are distinguished in Table 3.1.

Measurement terms are parameters of variables or constants which are either quantitative (e.g. counts, mass, length) or qualitative parameters (e.g. temperature degrees, coded shades of leaf colour). As criteria they require an explicit set of procedural rules and operational definition with appropriate methods and units of measurement. Units and dimensions have to be defined, for example, time duration [T], length [L], count [N] etc. Some concepts and their associated terms have dimensions, e.g. sporulation density $[N \times L^{-2}]$, apparent infection rate r $[T^{-1}]$, whilst ratios and efficiency terms have no dimension [1]. The weighting of parameters in the course of data evaluation should be based on a clear reference, as in Table 4.12, or the darkest normal green of leaves = 1 down to 0 for yellow or brown. Weights for variables and their parameters can also be obtained from statistical analysis, e.g. from multivariate regression analysis, principal component analysis or factor analysis.

Examples of measurement terms are presented in Table 3.2 along with their dimensions selected from Butt and Royle (1980, Tables 2–4) arranged in three classes I, II and III at the three levels, non-parasitic, parasitic and epidemic as in Table 3.1 with sublevels A, B and C at the parasitic level. Terms in class I are clearly and explicitly defined. They are practically equivalent to operational definitions as they provide rules for measurement. Concepts associated with terms in class III have a high degree of abstraction and some ambiguousness and often can only be measured with difficulty. Class II contains intermediate terms. Sub-level A at the parasitic level stands for inoculum terms, B for temporal terms and C for inoculum production terms.

Parameters as criteria are tractable as state or rate variables. They will be defined in the following chapters in relation to the relevant topic. Among the state variables, criteria with the highest degree of invariance or consistency in a unit under comparison, in general, have the highest discriminative value, e.g. in cluster analyses. Invariance of criteria is either due to low variability or a high degree of correlation. This correlation has to be defined quantitatively, even to allow for a defined reasonable variation, e.g. within confidence intervals (Madden 1986). However, redundant attributes and parameters common to all the units under study (e.g. airborne, leaf spot disease) should be excluded as criteria in a given study, e.g. when comparing airborne leaf spot diseases.

In conclusion, when descriptive and measurement terms are used as criteria in comparative studies the adoption of accepted epidemiological terminology is strongly recommended. Terms newly introduced should be clearly defined. For conformity of terms used in a report, a commonly and internationally accepted list of defined terms should be followed (and quoted).

For the proper use of terms, emphasis on the taxonomy of pathogens is appropriate. A wrong or outdated nomenclature can have an ill-effect on comparisons. For convenience older still and better known names may be given in brackets, e.g. *Stagonospora (Septoria) nodorum*. When the name of the teleomorph is quoted (e.g. *Phaeosphaeria (Leptosphaeria) nodorum*), the one of the anamorph should be given as well. This also applies to the subspecies level, i.e. variety (var.), pathovars, formae speciales (f. sp.) and even pathotypes (races). For fungi one could rely on the nomenclature adopted by the Commonwealth Mycological Institute, London, usually updated on CD-Rom. Modern molecular techniques like ELISA, PCR and similar ones may be used for identification purposes, even for rapid in-situ determinations.

Measurement terms may be subject to fuzziness. This can result from uncertainties in measurements (e.g. disease assessment), or data acquisition in general, from variability, heterogeneity, complexity in data sets and from scaling. This certainly is even more so with descriptive terms, particularly when based on subjective judgments or interpretations, vague factors and knowledge from different sources. The "soft computing" in fuzzy modelling (Bardossy and Duckstein 1995) is already employed in the development of decision-making aids, e.g. for integrated pest management (IPM) schemes (Apel et al. 1999). The utility of fuzzy methods and defuzzification (see Bardossy and Duckstein 1995) for comparative epidemiology still has to be explored. They may be of advantage to enhance results from comparison, for instance, in large-scale comparisons, e. g. geophytopathology (p. 154), geographic information service (GIS) applications (p. 155) and posterior analyses (Sects. 3.2.2.2 and 3.2.2.3) and for management models.

3.2 Data Acquisition for Comparative Epidemiology and Evaluation

Common experimental techniques and approaches, as used in plant pathology and ecology, are applicable to comparative research. Whenever possible, specially designed and appropriate field, growth chamber and simulation experiments would be most appropriate for comparative epidemiology. However, epidemiologists may not always have the resources for these often extensive, costly and complex field experiments. For this reason, in addition to experiments, posterior evaluation (Sect. 3.2.2) of available data could assist comparative epidemiology to achieve its full potential in across studies that have been published or in files for generalised concepts about epidemics and their elements and structures. The various sources for the data acquisition by experimental and posterior comparison are summarised in Table 3.3. Examples of the potential of both approaches, experiments and posterior evaluation, are presented in Chapters 4–7.

Comparison is easiest with the least ambiguous results at the same systems level, e.g. at the disease and epidemic levels. At the level of disease, component analysis of the monocycle of disease (e.g. slow rusting) and life table statistics (Zadoks and Schein 1980) would be the choice approach (Chap. 4). Comparative epidemiology at the more complicated community level is still little developed, although attempts have been made (Sect. 3.2.1). Examples of regional, national or even global comparison are, for instance, the composition of barley powdery mildew pathotypes in Europe (Limpert 1987) and research networks to compare the epidemic behaviour of pathosystems under different agricultural conditions (e.g. Royle et al. 1986). For comparisons at such higher levels, which may comprise several pathogens, varying conditions etc., posterior analysis could be considered, at least for cursory evaluations.

Table 3.3. Sources for data acquisition and evaluation in comparative epidemiology (with numbers of sections where dealt with)

Comparative experiments "across" pathosystems (Sect. 3.2.1)
  Field experiments (Sect. 3.2.1.1)
  Experiments under controlled conditions (Sect. 3.2.1.2)
  Comparative experiments by simulation (Sect. 3.2.1.3)
Posterior analyses across studies (Sect. 3.2.2)
  Classical reviews as qualitative summaries (Sect. 3.3.2.1)
  Quantitative summaries or collations of quantitative results (Sect. 3.2.2.2)
  Meta-Analyses: new evaluation of collected data sets from various primary studies (Sect. 3.2.2.3)
Evaluation procedures in comparative epidemiology (Sect. 3.3)
  Non-statistical comparison (Sect. 3.3.1)
  Overview of statistical methods (Sect. 3.3.2)

3.2.1 Comparative Experiments Across Pathosystems

In experimental approaches to comparative epidemiology at least epidemics (or some of their aspects) of two pathosystems should be investigated in the field or controlled environments with simultaneous measurement of relevant parameters. Simulation or model computation with validated models could be employed to substitute field experiments and test a variety of epidemiological aspects. These comparisons may be done on entire epidemics (disease progress curves or gradients), or on relevant attributes or parameters, e.g. initial disease intensity and its relationship with maximum disease intensity.

3.2.1.1 Field Experiments

Various types of field experiments, including holistic designs, with measurements of hosts, pathogens, diseases and climatic factors, are reviewed by Aust and Kranz (1988). When appropriate, concepts and methods adopted from ecology could also be employed. They are of use for comparative epidemiology as well, with crops in experimental plots and fields of any size. Appropriate scientific planning and conduct of experiments and their subsequent statistical analysis have, of course, to be observed (Kranz and Rotem 1988; Kranz 1990a).

The design of experiments for comparative epidemiology has to ensure strict comparative measurements of relevant parameters of pathosystems. These criteria will be used to compare structural elements, progress curves, interactions and factors that affect them. Peculiarities of the object, or topic, should be taken into account. For instance, in virus disease, vectors may be a component of the pathosystem. Therefore, the entomological aspects may require the cooperation of an entomologist (Thresh 1978, 1981, 1983; McLean et al. 1986; Raccah and Irwin 1988). Sources of cryptic errors (Van der Plank 1963) should be eliminated or stated if they were unavoidable, e.g. for the benefit of posterior analysis. Particular aspects of experimentation with soilborne and root diseases are dealt with in the volume by Campbell and Benson (1994a).

When comparisons of spatial aspects in field experiments are planned, the pattern, distribution and density of host plants should be taken into account when choosing the appropriate experimental design and sampling methods. Information from Nicot et al. (1984), Nicot and Rouse (1987), Madden and Hughes (1995b) and Hughes et al. (1997) may be considered in the planning process. When epidemics of the same pathosystems are to be compared between greenhouses and fields (Pataky et al. 1983; Jarvis 1989) both sites should be adjacent and the same cultivars cropped with all other treatments using the same equipment in either site. The effect of seasonal meteorological influences could be compared by means of staggered planting of the crop. Yang et al. (1990) did so with 73 sequential planting experiments of soybean rust epidemics. This ensured a broader data base for comparisons of seasonal effects on the disease.

Comparative field experiments in unrelated ecotopes, e.g. diseases of field crops and forest trees, various pathotypes, or even the same disease at different localities, climates, years and cultural practices should use as strictly as possible the same experimental methods and criteria in all these projects. This advice also applies to comparisons of epidemics that have different ecological or historical backgrounds (e.g. root and leaf diseases; endemic and invading pathogens). These comparisons may be more successful with closely related pathogen species (Weltzien 1972). Coordinated joint projects of several researchers may help to overcome existing difficulties. Such a collaborative programme was done on epidemic patterns of *Stagonospora nodorum* in western Europe from 1981 to 1983 (Royle et al. 1986). Other examples are the publications of Rudgard et al. (1993, see Sect. 6.4) and Savary et al. (2000a,b).

Experimental comparison across pathosystems in the field either for entire epidemics or some relevant aspects is possible for the following six cases 1–6:

1. concurrent comparison of various pathosystems in one population of a host plant species (e.g. the same crop), i.e. syn-epidemiology (see below);
2. the same pathosystem affecting different crops in one field (e.g. barley and wheat);
3. epidemics of various host species (plant communities, ·vegetation types) with one or more pathosystems at the same site and time, for instance, in mixed in situ wild plant communities, or as designed experiments (p. 19);
4. epidemics of one or several pathosystems at different localities, climates, years and cultural practices with strictly the same experimental design and methods if feasible;
5. comparisons among ecotopes (agro-ecosystem, regions): (a) same or different pathosystem, (b) various pathosystems in wild plant communities, (c) results from surveys;
6. comparisons of the behaviour of pathogen genotypes.

Comparative field experiments for 1. to 3. are more easily performed in one locality with two or more diseases of a particular crop, e.g. ear blotch, powdery mildew and stripe rust in wheat. Similarly, under the same edaphic and climatic conditions, the same diseases that affect different crops (i.e. different pathosystems) can be compared. Table 3.4 lists examples from experimental studies which contain disease progress curves, appropriate growth functions and disease assessment, various applications of epidemiological methods (e.g. sampling methods, crop loss estimates), spatial aspects and cultural practices.

Comparative field experiments, as in cases 4 and 5, may require cooperative research organised either as networks or, more tightly, in consortia. For these projects the same methods (down to the field record forms) should be employed as strictly as possible. In unrelated ecotopes such as field crops vs. forests, commonalities in criteria should be sought and agreed upon. Fewer, but more comparable criteria, might be a better start than a fully fledged attempt.

A syn-epidemiological study (see case 1, above) of practically all pests occurring in a crop is a special form of syn-ecology and of comparative epidemiology. Its major objective is to elucidate the strength of the joint effects of interaction and competition among the various associated pests on the intensity of diseases in the crop and their epidemics. Syn-epidemiological studies can cope with the complexities of all concurrent or successive pests in a field and compare their interactions and behaviour. Some results that emanate, e.g. multi-factorial loss functions and loss profiles (see Sect. 7.1), then become rational bases and functions for IPM schemes. Holistic experiments are appropriate for syn-epidemiological research in

Table 3.4. Selected examples of specific comparative field experiments

Objective of comparison	Authors
Epidemics studied in wild host plant population (pp. 19; 130)	Kranz (1968a–c); Frinking and Linders (1986)
Infection of wheat by Stagonospora nodorum and Septoria tritici	Cooke and Jones (1970)
Disease progress of *Mycosphaerella fijiensis* and *M. musicola* on *Musa* spp. (p. 69)	Mouliom-Pefoura et al. (1996)
Growth functions for progress curves (p. 34)	Analytis (1973)
Comparison of Gompertz and logistic functions (p. 110)	Plaut and Berger (1981)
Relationship disease severity viz. incidence of two diseases	Analytis and Kranz (1972)
Multivariate comparison of papaya ringspot epidemics (p. 36)	Mora-Aguilera et al. (1996)
Syn-epidemiological studies	Jörg et al. (1990)
Of the pest complexes in wheat (see below)	
(p. 161)	Wiese et al. (1984)
(p. 161)	Johnson et al. (1986)
(p. 36)	Khoury (1989)
Aspects of disease complexes in white clover (pp. 126)	Nelson and Campbell (1992, 1993a,b)
Effect of leaf wetness and temperature on disease intensity of four tomato diseases (p. 70)	Bashi and Rotem (1974)
Interplot interference	Paysour and Fry (1983)
Evaluation of field sampling techniques for two rice pathosystems	Disthaporn et al. (1993)
Epidemics in greenhouses, microplots and fields	Pataky et al. (1983); Jarvis (1989)
Effect of sowing date, cultivar and race on *Fusarium* wilt of chickpeas (p. 170)	Navas-Cortés et al. (1998)
Crop loss assessments models for several bean pathosystems (p. 162)	Schuld (1996)

which systems analysis, with the measurements of as many variables from the host plant, pests and meteorological factors as needed and feasible, can be used. The interactions observed between pests either were neutral, competitive or predisposing for other pests occurring in the crop.

Syn-epidemiological studies were conducted in replicates in farmers' fields of winter wheat with naturally occurring pests in the field and no control measures (Bonfig-Picard 1982; Bonfig-Picard and Kranz 1984; Jörg 1987; Khoury 1989; Kranz and Jörg 1989; Jörg et al. 1990; Khoury and Kranz 1994; Weber and Kranz 1994a,b; Weber 1996). In these experiments single plants or tillers and miniplots were the sampling units which provided a sufficient number of sample sizes to allow for the many replicates needed for statistical analysis. The variables and parameters of pest and host development are described in Table 3.5. Untagged plants were removed at random for destructive sampling to assay for hidden pests, besides those on the roots and stem bases.

The measurement terms in Table 3.5 resulted in large numbers of variables and complex data sets which required multivariate statistics for their analysis and evaluation (Sect. 3.3.2). Examples of results obtained by this comparative research are presented in Fig. 4.15 and Tables 4.21 and 5.15.

Pathosystems in wild host plant populations (case 3) are useful and convenient for comparative epidemiology (Kranz 1968a). Wild host plants as mixed species or pure species communities can be studied in replicated fixed plots of a reasonable size, or as single plants, marked within continuous populations. Relevant parameters are to be recorded for at least 2 years as in Table 3.5. In addition, Dinoor and Eshed (1984) and Burdon (1993) stress the utility of wild plant populations for studies of epidemiological aspects, particularly for the effects of resistance genes in a mix of different R-genotypes on long-term behaviour of diseases. Studies on wild host plants already have made an impact on control strategies with crop resistance, e.g. multi-lines and mixtures (Sect. 7.3). Wild host plant populations offer opportunities, particularly for long-term studies. For instance, the effect of external conditions that provoked a flare-up epidemic by disturbing the balance of usually low disease intensities (Kranz 1990b).

The advantage of epidemiological studies on wild host plants is the possibility of an immediate and concurrent comparison of epidemics of various pathosystems. All of them supposedly live under conditions suitable for them and they are exposed to the same microclimate at short distances. This permits the derivation of conclusions as to what extent facets of epidemics and factors that affect them, can be generalised and the differences that exist among pathosystems can be distinguished. These studies are rather inexpensive, but they mainly serve basic research. However, if pathogens are involved, or related ones which also affect crops, clues may by obtained as to the effect intensified agriculture practices have on their epidemics in crops.

Table 3.5. Design, methods and parameters of holistic field experiments in syn-epidemiological studies in winter wheat. (Adapted from Kranz and Jörg 1989)

Single tagged tillers: randomised, with sample sizes between 400 and 1500 main culms and, if necessary, secondary ones;

Separate tillers: untagged for destructive random sampling;

Miniplots: of 0.5 m^2 for measurements and with removable plastic cabins of 3×1.5 m for artificial infections;

Natural infection/infestation only, except in cabins;

Recordings: done in defined growth stages of wheat (DC-scale), usually 7–8 times from stage 37–77 DC;

Host parameters: growth stage, length of culms and leaves, leaf area per insertion, length of ear, number of grains/ear, grain weight, thousand kernel weight (TKW), "host morphology" (length of culm × diameter of culm × length of ear);

Pest parameters: incidence (%) and severity (%, or numbers) per leaf insertion;

Weather parameters: temperature, relative humidity and rain recorded continuously;

Fixed-formatted field record forms, filing on PC;

Supporting experiments: in phytotrons, growth chambers and greenhouses.

Wild plant communities were studied by Kranz (1968a–c) who used four sites as plots with mixed-species communities on the premises of an experimental station in Guinea, West Africa. Variables and parameters were recorded monthly for 2 years on the 50 plants species that occurred in at least two of these plots. These were the relative density and growth stages of host plants and both disease incidence and severity. The objective of the study was to compare disease progress curves on as many host species as possible under practically the same conditions and determine whether diseases were affected by host density and development. Some of the results are given in Fig. 3.6 and Tables 5.2–5.6. A similar approach was used for a comparison of the parasitic mycoflora in six types of vegetation on four different soil types (localities) around Giessen (Kranz and Knapp 1973; Fig. 6.8).

An experiment was designed and conducted for 3 years with five species of wild host plants and spring barley planted adjacently in large pots with ground contact (Kranz 1975a,b, 1976, 1977). The plants were exposed to the same climatic factors. Apart from host development (time of leaves unfolding and disappearing, their physiological state by colour and flowering) the appearance of lesions on each leaf, their number and size were counted and measured weekly on each leaf. The comparisons that resulted from the experiment are dealt with in Figs. 5.5 and 5.6 and Tables 5.9–5.11.

3.2.1.2 Experiments Under Controlled Conditions

In comparative research, experiments in growth chambers, phytotrons or other controlled conditions (see Rotem 1988) may be used to study: (1) components of the monocycle in epidemics (see Chap. 4), (2) the falsification of results from field experiments, (3) in-depth factor analysis, e.g. the effect of leaf wetness duration on sporulation, and (4) life table studies to elucidate the components of the monocycles of disease for their possible impact on, and implications for, the course of their epidemics, as demonstrated by Sache and Zadoks (1995).

Across studies under controlled conditions can also be done with polycyclic diseases, for instance, as mini-epidemics. Cohen and Rotem (1971) first used this approach to clarify aspects of the epidemiology of *Pseudoperonospora cubensis* on cucumbers. In their design of mini-epidemics they employed three chambers, each with a different cycled temperature regime and constant humidity (min. 55% relative humidity at the hottest time of the day) and 12 h of "dew" produced by humidifiers during the night. One-week-old potted cucumber plants were placed in the chambers. As the source of inoculum, one diseased plant was put into each chamber for 16 h (dew period for sporulation, plus 4 h for dispersal assisted by a blower in the cabin). The setup was then left alone until the recording of the results. A suction spore trap may be installed, e.g. to measure spore dissemination in the cabin. It is also possible to compare concurrently relevant components of several pathosystems in growth chambers, if space permits (Bashi and Rotem 1974, 1975). Table 3.6 lists some typical examples from research under controlled conditions for comparative epidemiology. These include comparative analysis of weather factors on infection and latent periods, the epidemics of pathosystems influenced by the host and an interacting virus disease, and the fitness of races. In a

Table 3.6. Selected examples of comparative experiments under controlled conditions

Effect of temperature regimes on infection and disease buildup of cucumber downy mildew in mini-epidemics (p. 19)	Cohen and Rotem (1971)
Leaf wetness duration affecting infection of four pathogens on tomato (pp. 19)	Bashi and Rotem (1974, 1975)
Temperature affecting latent periods of *Stagonospora nodorum* on wheat and *Erysiphe graminis* on barley (p. 68)	Aust and Hau (1983)
Epidemics of two pathosystems of *Leveillula taurica* on pepper and tomato (p. 60)	Reuveni and Rotem (1973)
Effects of a virus disease on two fungal diseases of beans at different temperatures (p. 73)	Bassanezi et al. (1998)
The fitness of two races of *Erysiphe graminis* f. sp. hordei over eight generations (p. 82)	Stähle et al. (1984)
Monocyclic components in a Life-table study faba bean rust	Sache and Zadoks (1995)
Climatic effects on monocyclic components of rust (*Puccinia sorghi*) and leaf blight (*Exserohilum turcicum*) (p. 73)	Vitti et al. (1995a)

comparison of monocyclic components of the pathosystems *Puccinia sorghi* and *Exserohilum turcicum* on maize, Vitti et al. (1995a) measured the parameters for disease efficiency, incubation period, rate of lesion expansion and spore production and factors that affect them.

3.2.1.3 Comparative Experiments by Simulation and Model Computation

Computer simulation (Waggoner 1978) is a versatile and powerful tool in comparative epidemiology, provided validated models exist. Simulation experiments are cheap, with no risk to crops and environment. They are flexible when used to compare the effects of any kind of known or assumed factors; parameter values on entire epidemics, or on their elements, structures etc. In simulators and models, parameters of state variables can thus be altered for comparisons with known or plausible values to be made, at least virtually. The effects that parameters have on disease progress curves, latent periods, asymptotes and the relation between the disease intensities y_0 and y_{max}, etc. can be examined. Simulators also allow the comparison of structural elements of epidemics. The results of these analyses can reveal the state variables that are most relevant as criteria for comparison and even their relative weight can be determined in a given system or context. Uncertainties can thus be clarified or assumptions falsified. However, "prior to assessing the simulated results of each comparative method, it is necessary to diagnose whether the simulations actually represent pre-established conditions" (Gittleman and Luh 1992). For posterior comparisons, sets of actual field data or available ones on file may be employed.

Elements and structures of simulation models (simulators) were compared by Hau (1985), Teng (1981, 1985) and Waggoner (1990). A general introduction to simulation is given by Waggoner (1974, 1978) and Campbell and Madden (1990). Rapilly (1991) presents a number of flowcharts and relational diagrammes and Table 3.7 lists some examples of comparison by means of simulators and model computations.

Once a valid simulator is available to mimic disease progress observed in the field, strategic options of chemical control can be compared as was done by Hau (1985) with GEMETA, his simulator of barley powdery mildew (Fig. 3.1). The efficacy of a fungicide in terms of elimination of 50, 70 and 95% conidia and the assumed lasting effectiveness of the fungicide on the plant can be shown. The curves

Table 3.7. Selected examples of epidemiological comparison by means of analytical models and simulators

Effect of parameters in Van der Plank's corrected basic infection rate R on epidemics (p. 65)	Zadoks (1971); Bergamin Filho and Amorim (1996)
Components of the infection chain of wheat stripe rust (p. 65)	Luo and Zeng (1995)
Sensitivity of estimates of parameters of growth functions (p. 125)	Thal et al. (1984)
Effect of y_{max} in functions to compute infection rates (p. 115)	Neher and Campbell (1992)
Effect of five epidemiological parameters on lesion expansion of *Cercospora medicaginis* on alfalfa and *Exoserohilum turcicum* on maize (p. 73)	Berger et al. (1997)
Comparison of three linked interaction models for epidemics of *Stagonospora nodorum* and *Erysiphe graminis* on wheat (p. 90)	Weber (1996)
Factors affecting endemicity (p. 95)	Onstad and Kornkven (1992); Jeger and van den Bosch (1994b)
Focal expansion of Puccinia striiformis and Peronospora farinosa (p. 140)	Van den Bosch et al. (1988a)
Stochastic spread in space and time (p. 139)	Xu and Ridout (1998, 2000)
Degree of hyperparasitation of *Puccinia recondita* by *Sphaerellopsis filum* and its effect on leaf rust epidemics (p. 23)	Hau and Kranz (1978)
Simulation of multiple pest damage in rice	Pinnschmidt et al. (1995a)
Effect of fungicide application on barley powdery mildew epidemics (Fig. 3.1)	Hau (1985)[a]
Selection of simple and complex races in relation to their initial frequencies using EPIMUL[b] (p. 24)	Lannou and Mundt (1996)
Effect of genetic, chemical and cultural control on epidemics of diseases caused by seven pathogens (p. 126)	Gilligan (1990b)

[a]A one-season simulation by means of GEMETA, an "open" simulator computing with real-time field data of an actual epidemic was used for this comparison.
[b]EPIMUL by Kampmeijer and Zadoks (1977).

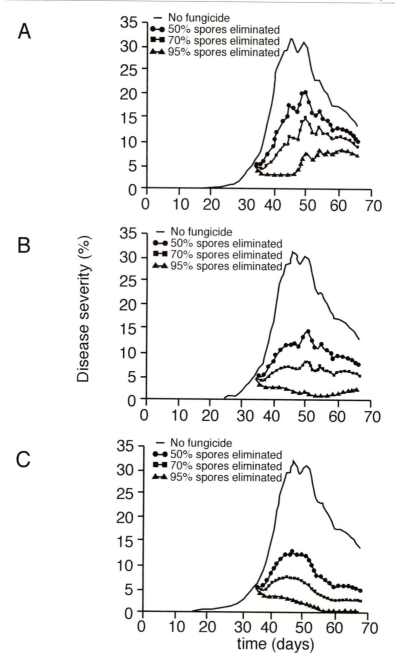

Fig. 3.1. A comparison of the effect of a fungicide on powdery mildew of barley (*Erysiphe graminis* f. sp. *hordei*) with three levels of inoculum reduced by 50, 70 and 95% and an efficacy for **A** lasting for 7 days, **B** 14 days and **C** 21 days compared to untreated. A data set of an untreated epidemic (Hau pers. comm.) was used for this simulation

indicate that Hau found that a short period of fungicidal activity on the host would not be enough to control powdery mildew. The longer the efficacy of the compound lasts, the more control was achieved in terms of a reduced disease severity during the epidemic. However, the difference between 70% of inoculum destroyed (probably more realistic) and 95% is not too great. Other considerations need to be examined, e.g. whether the remaining disease severity exceeds the physiological damage threshold and questions of control cost/return.

In the course of the validation, simulators may fail to produce an acceptable fit with observed data plots of the pathosystem for which it was developed when tested under different environments or agricultural practices. This is not necessarily an indication of insufficiencies in the simulator. It could well be due to different conditions affecting the pathosystem somewhere else. Such a response was experienced by Amorim et al. (1995) when they applied their simulator (Berger et al. 1995) to observed disease progress. The parameter values were valid for epidemics of bean rust in one Brazilian state, but not in another state where a different climate prevailed. Of course, if the simulator is to be generalised, the functions with their parameter values will have to be adjusted. In the context of comparative epidemiology, this example of the different behaviour, due to environmental factors, of the bean rust pathosystem at the two localities shows the necessity of an across-study on the climatic factors at both sites.

Analytical models sensu Jeger (1986) can, of course, also be employed for comparative epidemiology by changing parameter values in the equations (Hau 1990) to compare these effects. These models may have a higher mathematical sophistication, be simpler in their structure than simulation models, with fewer variables and parameters and, thus, less input data (Jeger 1986). However, analytical models are often limited to a single output, e.g. the effect of the corrected basic infection rate R on the epidemic (Van der Plank 1963; Vanderplank 1982). Analytical models tend to be more lucid and, therefore, are more appealing.

For instance, Vanderplank (1982), Jeger (1984) and Jeger and van den Bosch (1994b) prefer analytical models to simulation. However, when these simpler models are transformed into computer models, as Zadoks (1971) did with Van der Plank's (1963) equation for $R = r[y_t/(y_{t-p} - y_{t-i-p})]$ with its parameters initial disease intensity y_0 (y_t in the equation), latent period p and the infectious period i over time, then analytical models become "closed" simulators. Closed simulators are operated by the researcher who enters parameter values from actual or historical field or weather data in accordance with his study objective. With such a simulator the effect of the hyperparasite, *Eudarluca caricis* (anamorph *Sphaerellopsis (Darluca) filum*), on disease progress of wheat leaf rust (*Puccinia recondita*) was simulated (Hau and Kranz 1978). The simulator was based on known disease progress curves of the rust and experimental data about the disease efficiency of *E. caricis* (i.e. the degree of rust pustules covered by pycnidia of the hyperparasite). The simulated components were: time of onset of hyperparasitation, proportion of pustules infected by *E. caricis* and percent coverage of pustules by pycnidia of the hyperparasite reducing sporulation of the rust. The model indicated that there can be a considerable efficacy of the hyperparasite in reducing rust severity up to about 90% and the disease progress of leaf rust.

Spatial aspects were first simulated by another closed simulator, EPIMUL (Kampmeijer and Zadoks 1977). EPIMUL was later found applied in similar theoretical studies. With a simulator based on EPIMUL Lannou and Mundt (1996) compared selection for simple or complex races in a pathogen population. Analytical models are nowadays widely employed in the comparison of spatial epidemic aspects. Jeger and van den Bosch (1994b) advocated the inclusion of stochastic effects in analytical modelling. Hughes et al. (1997) reviewed, in some detail, modelling and model validation for spatial patterns of disease incidence.

3.2.2 Posterior Analysis Across Studies

Although field experiments would be the most desirable approach to achieve the objectives of comparative epidemiology, they have their limitations (Sect. 3.2.1). In addition, the validation of results from field experiments can be demanding and may pose problems as they may require considerable resources of manpower and funds and time to complete such projects. Thus, comparative epidemiology faces a dilemma: there are only limited resources for appropriate experiments "across". However, there are abundant published data from experiments with single pathosystems or other related experiments which could be used for posterior quantitative analysis. The latter would be a shortcut to enhance epidemiological insights. The posterior analysis of studies across pathosystems is thus more than a supplementary approach of comparative epidemiology to utilise existing data, information and material. Published or unpublished results or data sets, properly used, can be used to filter, summarise or generalise relevant information on epidemiological aspects and uncover challenging views behind the multitude of apparent diversity. Posterior analysis may even be the last resort for comparisons of topics in unrelated ecotopes using data sets from field experiments obtained at different sites and times.

It appears that many issues of practical relevance could be investigated by posterior analysis. For instance, which weather and host constellations in a season (or during overseasoning) are conducive to years of severe intensity of powdery mildew, rust or a leaf spot disease of a crop? What are the effects of cultivars and pathotypes on the epidemics of a certain disease? Comparison by means of posterior studies across different conditions can assist in the development of more generally applicable forecasts, or consolidated functions for crop loss assessment (e.g. Waggoner and Berger 1987) for a given disease etc. (Sect. 7.1).

In medical research, limitations to research have prompted the use of posterior evaluation. Among the major motives are (Blettner et al. 1997): improvement of estimates of effects of factors, qualitative and quantitative risk assessment, reasons for heterogeneity of results, analysis of dose-effect relationships, the enhancement of the statistical power of available data and better estimate of effects. Also, in plant disease epidemiology posterior analyses are not uncommon (Table 3.8).

Posterior analyses require a suitable database of individual studies consistent with the research objective and with the standards of experimental research. Methods, parameters and measurements that are utilised should be appropriately comparable among the data sets used from different sources with their specific characteristics for the study and be sufficiently homogeneous. If heterogeneity exists among the data sets, this should be clarified before the pooled data are analysed to avoid (or explain) possible erratic effects in the results and subsequent validation. In any case, caution and sensitivity to the use of data, the interpretation

Table 3.8. Some examples of quantitative posterior analyses in plant disease epidemiology

Topic	Number of studies[a]	Conclusions	Authors
Effect of resistance on disease progress (p. 34)	117	Difference between vertical and horizontal resistance	Van der Plank (1963)
Gradients (p. 144)	124	Frequency of gradients following exponential and power laws	Gregory (1968)
Gradients (p. 144)	105	Point source favours exponential and line source the power law	Ferrandino (1996)
Frequency distribution of lesions on plants	112	Most lesions are not randomly distributed	Waggoner and Rich (1981)
Lesion expansion simulated (p. 73)	86	Lesion expansion contributes most to disease severity	Berger et al. (1997)
Effect of farming practices: organic, integrated, conventional (p. 169)	38[b]	Little differences in disease severity among practices[c]	Van Bruggen (1995)
Effect of control measures on parameters of DPC (pp. 31, 124)	33[d]	Reducing the asymptote and delaying onset of epidemic were most effective	Gilligan (1990b)
Disease control by mixtures in barley (p. 173)	11	Comparing yield in mixtures and their components	Wolfe and Finckh (1997)
Crop loss assessment for *Cercospora beticola* on sugarbeet[e] (p. 32)	113	Derivation of equations from field data of 9 years with multivariate regression analysis	Kelber (1977)

[a]Number of published studies evaluated.
[b]Number of pathosystems × sites. The disease severity was rated in nominal classes 0, 1, 2 and 3.
[c]There is a certain tendency for higher severity ratings under conventional practices, e.g. powdery mildew and eyespot of cereals, the only case listed of potato late blight was most severe under organic practices.
[d]Seven pathosystems each with one to several treatments per disease evaluated.
[e]From this posterior evaluation three multivariate regression functions emerged with four parameters each and R^2 between 0.83 and 0.88.

of published information and reasoning behind arguments by the authors of the studies are necessary (Gittleman and Luh 1992). Whenever possible, conclusions derived from posterior evaluation of evidence obtained from across studies should later be validated experimentally, either in field experiments with standardised layouts (e.g. randomised block design), under controlled experimental conditions or by means of computer simulation.

Among the posterior analyses Blettner et al. (1997) distinguish four types: (1) classical reviews as qualitative summaries, (2) quantitative summaries, (3) meta-analyses as a new evaluation across available studies and data sets (Sect. 3.2.2.1–3.2.2.3), and (4) prospective planning for pooled evaluation. Strictly speaking, type (4) is not a posterior analysis (Blettner et al. 1997), but a coordinated research project, organised either as a loose network or as a consortium. In consortia, partners share the work of a programme, for instance, on the epidemics of the same pathogen under different conditions, localities etc. with appropriate methods and the scientists publish jointly.

3.2.2.1 *Classical Reviews as Qualitative Summaries*

These reviews (type 1 of Blettner et al.) usually have limited scope or relevance if they are subjective, unreliable, unsystematic or incomplete. They may also be bias-prone due to the selection of studies included to back a preconceived concept. Reviews of this type often accumulate and filter facts but do not always reach definite conclusions. However, even imperfect, inadequate or biased reviews, if done as carefully and as objectively as possible, at least reflect the state of research. Reviews of types 1 and 2 could be improved by evidence from CD-ROM literature retrieval or information from the Internet. A nearly complete selection of papers for a review of a topic is no longer a problem nowadays. Papers published in languages other than English usually have English summaries. Before conflicting results are excluded, one should first check whether they are from sampling errors, differences of measurement and computational or clerical errors (Table 3.9). These differences may be corrected. However, genuine conflicting results should be explicitly dealt with. Hunter and Schmidt (1990) suggest that reviews should be done according to an agreed protocol which ensure that (1) information on the process of the literature search is given, and (2) criteria applied for the exclusion of published evidence are stated. Finally, (3) whenever possible, magnitudes of the primary findings such as means and variances and other major characteristics should be presented, or kept, particularly for type 2.

Some reviews were specifically written to extract general conclusions from scattered evidence of comparative epidemiology, for example on: methodology (Kranz 1988b), temporal patterns of fungal diseases (Cox and Large 1960; Merrill 1967; Kranz 1974a; Croxall and Smith 1976; Waggoner 1986; Gilligan 1990b), gradients of diseases (Wolfenbarger 1959; Gregory 1968; Fitt and McCarthney 1986; Minogue 1986) and of virus spread (Thresh 1974a, 1978, 1983). In addition, reviews that do not originate from specific comparisons, e.g. pure descriptions of phenomena, reports and reviews, may yield useful information, like some of the papers in the volume edited by Garreth-Jones (1998). Information from such publications may contain new facts and they may incite intuition, speculation and ideas which give rise to new research hypotheses. Reviewers may uncover dis-agreement between accepted theories. Good examples of this, though not specifi-cally written for comparative epidemiology, are the monographs by Robinson (1976) on pathosystems and by Palti (1981) on cultural control of diseases. Palti has summarised available information on a number of aspects of cultural control in tables, thus comparing effects. The chapter by Thresh (1983) is in between qualitative and quantitative review. He interprets graphics of disease progress curves of virus diseases, discusses the factors affecting them and their implications for disease control.

Qualitative reviews of relevant information are particularly useful if the infor-mation from the literature is enriched by the reviewer's own research. Examples are the studies by Zadoks (1961) on stripe rust of wheat (*Puccinia striiformis*) and Populer (1972) on powdery mildew of rubber (*Oidium heveae*). Both authors identified particular traits of these diseases, explained some of the sources of variation in the behaviour of their epidemics under different conditions and sug-gested theories. Such evaluations are usually done critically and competently, al-though some bias may remain.

3.2.2.2 *Quantitative Summaries of Published Data*

Type 2 posterior evaluation uses quantitative data such as published tables and graphs. The quality of this type of summary depends very much on the availability of quantitative evidence, e.g. tabulation of statistically significant or non-significant results that may include, for instance, correlation matrices to permit re-calculation of regressions etc. Quantitative reviews still may be as biased as qualitative ones, particularly when selected and explorative data sets are reviewed. Major problems with the available studies are differences in the research plans, methods of measurement, definition, use of independent variables and criteria and methods of statistical evaluation. It may be recommendable to contact authors for relevant, but unpublished information. Posterior evaluation of quantitative infor-mation from published evidence certainly would benefit from correlation matrices which usually are not published in papers. As journals usually do not print such basic information, researchers could preserve the matrices and make them avail-able on request for posterior analysis. A comparison of graphs of disease progress curves, e.g. for epidemics of potato late blight (*Phytophthora infestans*) by Cox and Large (1960) and Croxall and Smith (1976), as well as gradients, is a quanti-tative posterior analysis. Some examples of quantitative reviews across studies in epidemiology are given in Table 3.8.

Type 2 may already come close to meta-analysis (type 3), which summarises information and statistically re-analyses results from individual studies in order to integrate the findings, evaluate data which otherwise may not be utilised to their maximum extent and identify better estimators for quantitative effects (Dickersin and Berlin 1992).

3.2.2.3 Meta-analysis, Evaluation of Data Sets
from Various Primary Studies

Type 3 posterior evaluation across studies is meta-analysis (MA) proper (Blettner et al. 1997). Pearson (1904, cited by Dickersin and Berlin 1992) was apparently the first who did a meta-analysis by averaging five estimates from samples of a population. Meta-analysis begins with data sets of studies suitable to provide empirical evidence for a topic and then extracts and codes information. Meta-analysis does not analyse studies, but study results, e.g. numbers, correlation coefficients or the distances between sample means, which can be compared, averaged, empirically tested for a hypothesis or analysed as long as there is no logical contradiction. This posterior analysis uses pre-specified variables (e.g. primary disease intensity y_0 and its effect on maximum disease intensity y_{max}) from individual studies, estimates their associations, sums up and re-evaluates the statistical evidence on "the effect measure" from each study included in a MA (Dickersin and Berlin 1992). For Hunter and Schmidt (1990) meta-analysis by data pooling and synthesis provides "...the empirical building blocks for theory. ...(its) findings tells us what it is that needs to be explained by the theory". Meta-analysis primarily is an explorative research tool that may not necessarily reach definite conclusions. Still, new hypotheses may emerge, and common estimators of effects with higher relevance can be expected than from type 2 (Blettner et al. 1997).

Meta-analysis as a research procedure still is not yet universally accepted. However, there are some obvious advantages of this procedure which need to be developed. One benefit from meta-analysis is the identification of weaknesses in research approaches, by which it may find and support new research needs or projects. For the first inroads an approximate answer from meta-analysis to the right question may be better than nothing. Meta-analysis may at least improve the quality of published and unpublished studies on certain facets in epidemiology. Topically oriented meta-analysis could prepare heterogeneous data sets to make them (more) comparable by eliminating study imperfections or man-made errors (Table 3.9). However, doubts about the quality and appropriateness of studies included in a meta analysis can be checked by formal tests of heterogeneity. On the other hand, if scientific rigor is maintained in MA one would, over time, expect consistent and stable results (Dickersin and Berlin 1992). Dickersin and Berlin (1992) feel that a too narrow, rule-oriented start of posterior across-studies may even be inconsistent with the creative nature of research. However, because of potential internal and external threats to the validity of the within-subjects design, meta-analysis may become misleading when its domain is stretched too far. The available data sets may lead to a bias which, however, is a rather general problem in research and not specific to MA. Methods exist to eliminate a bias.

Nevertheless, practical needs to consolidate available knowledge favour the use of meta-analysis, also for the furtherance of a general baseline information in plant disease epidemiology. In medical science, for instance, it is already one of the methods to establish "evidence-based medicine", e.g. in the "Cochran Collaboration Centres" as a more reliable computerised database for diagnoses and treatments. Meta-analysis also finds application in medical epidemiology, pharmacy, social sciences and psychology and in industrial-organisational psychology. Even

Table 3.9. Sources of errors (study artefacts) in meta-analysis that may affect the value of output obtained from a posterior evaluation across studies (adapted from Hunter and Schmidt 1990)

1	Sampling errors
2	Errors of measurement of dependent and independent variables
3	Errors from dichotomisation of dependent and independent variables, e.g. when dichomised "more than" versus "less than" in dependent and "acceptable" versus "reject" in the independent variable
4	Errors due to range variation in the variables, when there is a systematic omission of parameters with values higher or lower than preconceived in dependent variables and data acquisition is confined to those with lower variation in independent variables
5	Errors caused by deviations from perfect construct validity in both variables. The study validity will differ from the true validity if the criteria of dependent variables are deficient or intercorrelated, or if the factor structure of the test differs from the usual structure for the same parameter or trait.
6	Reporting or transcriptional errors, viz. inaccuracies in coding data, computational errors, errors in reading computer outputs, typographical errors. These errors may be important
7	Errors due to variance caused by extraneous factors, e.g. a study included in the MA, differs to a great extent in objective, factors affecting parameters, conduct of research etc.

if no practical need favours the use of meta-analysis its results still may reflect the state of research and help to formulate new hypotheses.

Petraitis (1998) discussed the use of meta-analysis in ecology as a way to rationalise comparisons across experiments. According to Petraitis, the method typifies the problems of scaling the variables and the meaning of their relative importance as it uses results from different experiments. Meta-analysis can standardise the responses in these experiments, which then can be compared on a single scale using standard statistics for their evaluation, like the t-statistics. However, the use of data sets with data measured at varying ranges of scales may introduce a degree of fuzziness into the analysis (p. 13). This is a general problem for comparisons across pathosystems etc. when information and data are derived, for instance, from maps with their symbols for features covering wide ranges of classification. This reservation can also apply for methods like GIS.

Dickersin and Berlin (1992) stress that the standard scientific rules for planning, data collection, analysis and reporting also apply to meta-analysis. Selected data sets should be of an acceptable quality on which the quality of the results very much depends. A project of meta-analysis can be rather demanding if original data coming from various experimenters are to be analysed. Contact or even co-operation with the original authors may help to overcome shortcomings, uncertainties and bias.

An important step in meta-analysis is the computation of certain critical estimators, e.g. correlations, variances of correlation (the average squared deviation from the mean) and the variance of sampling errors. The question that arises in any across-study is how much weight to give to each individual result and what weight should be used once artefacts (Table 3.9) are corrected. If any substantial variation across studies occurs, then theoretically plausible independent variables

in each of the studies should be screened for their possible impact on these variations, or the authors consulted. After such clarification the meta-analysis concept,
according to Hunter and Schmidt (1990), is a simple process following the procedures (1) to (4) and appropriately adjusted to the research objective:

1. Calculate the desired descriptive statistics for each study available and average
   statistics across studies, e.g. correlation coefficients.
2. Calculate the variance of the statistics across studies.
3. If necessary, correct the mean and variance of study inaccuracies (artefacts
   other than sampling error; Table 3.9) by subtracting their amount. A bias still
   may exist because of differences in reliability of data or parameter values,
   range of data, validity of data construct, etc.
4. Compare the corrected standard deviation to the mean to assess the size of the
   potential variation in results across studies in qualitative terms. If the mean is
   more than two standard deviations larger than zero, it is reasonable to conclude
   that the relationship is always positive.

For the correction of errors in Table 3.9 variance across studies due to artefacts
and variance across studies due to real independent variables (e.g. disease intensity, temperature) have to be distinguished. If the data are homogeneous, simple
averages and variances could be computed. If such an homogeneous data set is not
available for independent variables, like disease intensity, temperature, etc., then a
test for heterogeneity is recommended.

Statistical methods in use in medicine and in social sciences (Hunter and
Schmidt 1990; Dickersin and Berlin 1992; Blettner et al. 1997) are: means and differences between means, weighted averages (e.g. the Mantel-Haenszel method for
odds ratios) either as constants, as functions of the sample size or within-study
variances (though often difficult with published results), correlation coefficients r,
contingency tables with the Chi^2 test, simple and multiple regression functions
and the confidence interval CI. The fixed-effects model (FE) and random effect
model (RE) may be used to cope with the variability among the studies, even
when heterogeneity exists, though there are some critical aspects with weighting
and interpretation. The "fixed-effects model" (by analogy with analysis of variance) assumes the existence of a single effect common to all studies, thus ignoring
the variability among studies in summarising the relative risks and odds ratios.
"Random effect models" rather tend to broaden the 95% confidence interval (CI).

Evaluations made solely by contingency tables and the Chi^2 test may give un-
adjusted relative estimates only and results may be distorted in the presence of
important factors as confounders. Also, the F-tests from individual studies in
meta-analysis can be a problem in cases of heterogeneity in the data set. Both tests
may uncover the sources of such heterogeneity, help to detect associations between characteristics and the possible outcome of studies, as long as the means are
not too different (Dickersin and Berlin 1992). Heterogeneity commonly is assessed by means of graphic methods, e.g. confidence interval (for time and space
related studies), scatter diagrammes, residuals. Heterogeneity affects the validity
of results with pooled estimators and the risk of wrong conclusions can be high.
Blettner et al. (1997) recommend sensitivity tests to enhance the stability of the
estimator. Sensitivity tests check the effect certain results have on the outcome of
a meta-analysis by omission of individual entries and recalculation of the MA

without the figure. For details of these statistical procedures and equations, we refer to Hunter and Schmidt (1990) and Dickersin and Berlin (1992).

In order to reap information from clinical and epidemiological research in medicine, Blettner et al. (1997), who also offer extensive references, should be consulted, or any other treatise on related statistics, e.g. Fleiss (1993). A common analytical theme in the techniques used to combine results across studies is the weighting average of study-specific results which vary from method to method. Blettner et al. (1997) point out that the methods in use do not always take into account the specific character of epidemiological observations in medicine. Risks in the use of standard methods of meta-analysis in the presence of heterogeneity in medical, therapeutical studies are reviewed by Ziegler and Victor (1999).

As has been mentioned, some sort of meta-analysis has already been applied in plant disease epidemiology. Gilligan (1990b), by means of the logistic function, summarised 33 disease progress curves of four polycyclic and two monocyclic pathosystems to test the effect of various control measures (Fig. 5.15). Neher and Campbell (1992) proved that infection rates are affected by the value for the possible maximum disease intensity in a pathosystem used in the equation. These papers and one, among others, by Kelber (1977) indicate that a clear objective is necessary, data selection has to be commensurate and strict, and appropriate statistical methods have to be employed. Other applications, for instance, could be validity tests of the relationship between the primary disease intensity y_0 and y_{max} across pathosystems. Also, the many crop loss functions (Sect. 7.1) published could be validated and, if possible, consolidated for a given pathosystem across cultivars, sites, years, agronomic practices etc. An example of such an attempt with "historical" data of the pathosystem *Cercospora beticola*/sugar beet is presented in Fig. 3.2 and Table 3.10.

Table 3.10. Validity tests of observed mean losses in purified sugar and the estimated values by the multivariate function (Fig. 3.2) in (%) for the pathosystem *Cercospora beticola*/sugar beet. (Kelber 1977)

Year of assessment	Loss in purified sugar (%)	
	Observed	Estimated
1954	55.2	55.0
	45.0	42.4
1955	23.2	21.5
	26.9	24.0
	29.4	28.2
	28.7	24.8
	37.4	39.7
	20.6	10.2
1956	26.8	24.4
1958	27.7	23.4
	32.0	38.5
	23.8	18.1
	24.3	18.0
1961	40.0	41.4
	35.4	36.8
1962	23.1	22.3

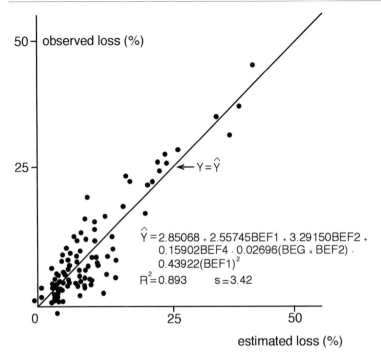

Fig. 3.2. Estimated loss of purified sugar caused by *Cercospora beticola* on sugar beet related to assessed yield loss by the function shown in the figure (both in %) with the variables BEG = disease severity (%) at the beginning of the epidemic in days after sowing,
BEF1,BEF2 and BEF4 = disease severity (%) at the end of the first, second and fourth
quarter of the progress curves. (Kelber 1977)

Kelber (1977) in his study on "historical" data sets used records on crop loss assessed between 1952 and 1962 from fungicide trials against leaf spot of sugar beet
on various cultivars at different sites around Regensburg. Altogether he had 113
progress curves of epidemics. From these data sets, those sets were selected which
were obtained with comparable methods of disease and yield assessment. Mean
values of four to eight replicated plots per year and site were used for analysis by
multivariate linear regression procedures. Various functions were obtained and
tested for their correct estimation. The one presented in Fig. 3.2 eventually turned
out to yield the most precise estimates of purified sugar lost in relation to yield
loss assessed in the field. This function was validated in Table 3.10 by omitting
the values of the particular year against which the correct estimate of loss in purified sugar is checked. The results, with few exceptions, are satisfactory. For the
years 1955 and 1958, other functions were somewhat better, but in turn, they were
inadequate for the other years. This indicates that more suitable data sets would
perhaps improve the function in Fig. 3.2 to be valid for all years.

3.3 Evaluation Procedures in Comparative Epidemiology

3.3.1 Non-statistical Comparisons

Comparative epidemiology of plant diseases also has become increasingly statistical. However, there are approaches for straightforward comparisons. These are deductions, visual comparisons and the compliance with existing concepts. These three approaches of comparison may be the initial stages to statistical analysis, or at least, to identify objectives for comparison in experiments or posterior analysis.

A single experimental result, observation or idea can induce a hypothesis. Zadoks (1978) discusses induction in some detail and its role in epidemiological research. However, many results, observations or experience rather lead to hypotheses through *deductions*. Although often non-systematic, deductions nevertheless have advanced useful notions, concepts and hypotheses, for instance, by Gäumann (1951) and Van der Plank (1963), the founders of modern plant disease epidemiology. Gäumann (1951) presented the first comprehensive, thorough, descriptive treatment of plant disease epidemiology and arrived at a number of now commonly accepted concepts, e.g. types of epidemics and principles like the infection chain. Among epidemics he typified explosive versus tardive epidemics, annual and secular or perennial (polyetic epidemics sensu Zadoks and Schein 1979), endemics (Sect. 5.1) and pandemics. From a geobotanic context he derived the concept of "re-encounter" and "new encounter" of pathogens and hosts and elaborated on the epidemiological implications of such events.

Van der Plank (1963, 1975; Vanderplank 1982) introduced mathematics to epidemiology and contributed further basic concepts. For example, the differential-difference equations to describe epidemic growth are now widely used to model epidemics. Van der Plank (1963) also deduced the various types of infection rates, descriptive terms like "simple interest disease", "compound interest disease", threshold theorem, cryptic error and endemicity (Sect. 5.1). His concept of focal and general epidemics (Sect. 6.2) describes spatio-temporal relationships in epidemics. By deduction, he also conceived the epidemiological implications of vertical and horizontal resistance. Van der Plank arrived at most of his concepts and models from observations and the reading of staff reports when he was director of a research institute. Typical is how he gleaned the concept of "Vertifolia effect" from a long list of results of cultivar trials on resistance against potato late blight (Kirste 1958). He mentioned that he was struck when he was reading the publication. This episode demonstrates how hypotheses can emanate from deductions which, in turn, can prompt experiments and possibly theories or principles.

Beyond mere deductions, comparison by *visual resemblance* can be done with simple plotting of disease progress curves etc., mapping or tabulation of quantitative data. A classical example of such a comparison in epidemiology are the disease progress curves which Van der Plank (1963) used to demonstrate the rate-reducing effect of non-race-specific horizontal resistance (Fig. 3.3) which has become a principle. Disease progress curves of virus diseases and effects upon their epidemics were also extensively compared visually, i.e. "biologically", by Thresh (1983).

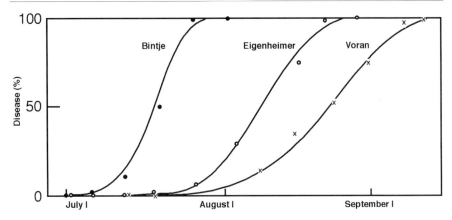

Fig. 3.3. Potato cultivars with no vertical resistance genes (*Bintje*) and vertical resistance plus different levels of horizontal resistance to late blight (*Eigenheimer* less than *Voran*) in 117 fields. (Van der Plank 1963)

Obviously, purely visual comparison has its pitfalls and may be misleading. Validation of the hypotheses, like the one in Fig. 3.3, can be done through specific experiments, posterior analysis or by statistical methods to avoid wrong conclusions. For disease progress curves and gradients, mathematical functions for their approximation are in use. Disease progress curves and gradients then may comply with the definition and be named by the function, e.g. an epidemic of logistic progress, or a gradient obeying the power law.

Compliance with a definition is instrumental for comparison as it decides whether a certain phenomenon complies with a theory or not (e.g. the rare black swan). In epidemiology, whether a disease obeys the definition of mono- or polycyclic can be established by the biology, e.g. the life cycle (Sect. 4.2), of its pathogen. Different parameter values of the components in the infection cycles and the resulting infection rates define whether a pathogen is an r- or K-strategist in ecological terms (Fig. 4.3). Rusts and mildews with their high rates rather behave like r-strategists with large fluctuations in population size. K-strategists, such as root diseases of trees, have low rates, but maintain a stable population size over time adjusted to a carrying capacity. Further examples of "compliance with a definition" are presented, amongst others, in Tables 3.1, 3.2, 5.13 and 7.3.

An important aspect of compliance are suitable functions to approximate and describe disease progress curves and gradients mathematically. Among the functions that fit observed disease progress curves (DPCs) of diseases, it would be of interest to know which function is the more common one for a given pathosystem. For apple scab (*Venturia inaequalis*), as an example, growth functions fitting the observed course of epidemics best were the Bertalanffy and the Gompertz functions (Analytis 1973).

3.3.2 Overview of Statistical Methods for Comparative Epidemiology

This section is a brief introduction to statistical methods which are also suitable to evaluate results from across-studies in experiments and posterior analysis (Sect. 3.2). Detailed information on suitable procedures can be found in statistical textbooks, handbooks for the appropriate computer software and in more recent reviews on statistics. There are appropriate statistical methods for ordination (Sect. 3.3.2.1) and classification (Sect. 3.3.2.3) introduced by epidemiologists in a number of chapters in Kranz (1990a) and summarised by Campbell (1998). An outline on statistical equivalence (Sect. 3.3.2.2) is added. The choice of a statistical technique for comparative epidemiology with their respective test parameters already has to be considered when a research project is planned depending on its objective and the type of data expected.

3.3.2.1 Methods for Statistical Ordination

Most methods of statistical ordination (Hau and Kranz 1990) can also be useful for studies across pathosystems when appropriately employed. These include means and standard deviation, correlation and regression analysis, analysis of variance (ANOVA), various multivariate methods like multivariate analysis of variance (MANOVA), multivariate regression analyses (MVR), principal component analyses (PCA), canonical correlation analyses (CCA) etc. Multivariate methods are most appropriate for the evaluation of holistic experiments though the variables measured are not always independent of each other and their data are not necessarily normally distributed. Many of the assumptions, procedures and applications of comparative methods are similar to problems in the time series literature (Cliff and Ord 1981). In any case, appropriate transformations for collinearity, normality and homoscedasticity may be required before analysis. When across-studies are planned, the qualitative and quantitative terms and scales of measurement should be applicable to all the pathosystems etc. in the study, either experimental or posterior.

After transformations, a sequence of statistical procedures may be necessary to do the comparisons. For instance, when Jörg (1987) analysed results from his synepidemiological field studies (Sect. 3.2.1.1) he started with the principal component analysis (PCA). With this procedure, experimental errors and multicollinearity, and thus, the number of variables to those of the highest effect were reduced. These variables then became the input for a canonical correlation analysis (CCA). Jörg (1987) replaced canonical weights by structural correlation coefficients, which show less variation and permit more unambiguous interpretation of the canonical variates. The analysis of the effect which the 8–15 pathosystems (diseases and pests) occurring during a season had on the yield of the wheat plants were submitted to multivariate regression analysis. Path correlation analysis computed the weights of the interacting individual pests together with the host parameter "morphology" (defined in Table 3.5) on yield (Fig. 4.14). For more details see Jörg (1987), Kranz and Jörg (1989) and Jörg et al. (1990). An even more sophisticated path correlation analysis, LISREL (**L**inear **S**tructural **Rel**ationship), was adopted by Khoury (1989) for her analysis of weather effects on several barley and wheat diseases (Tables 5.14, 5.15).

Uni- and multivariate regression analyses are commonly used in epidemiology. However, they may not be free of danger in their interpretation. Petraitis (1998) discusses some problems of inferring magnitudes of effects and importance because of the influence exerted on the amount of explained variance (e.g. R^2) by scaling, standardisation and covariance. The interpretation of regression coefficients depends on the range of variation of the independent variable, the standardisation of the independent variable in multivariate regression analyses with parameters of different metrics (number, length etc.) does not always "put them on equal footing", and finally, the effect that covariation can have on the regression coefficient. To obviate these sources of misinterpretation Petraitis (1998) suggested that when a study is designed, the setting of baseline data should be in a meaningful (epidemiological) context to prevent inferences about the magnitude of effects "outside a specific experiment".

PCA is a multivariate procedure to reduce complex relationships in observed data into simpler forms through a reduction in dimensionality of an original set of correlated variables (Campbell 1998) which is transformed into a new set of uncorrelated variables. These are called principal components (PC), in which the first few components retain most of the information of the original data set. The reduction in dimensionality is achieved as the routine assigns weights to each variable and then computes the corresponding linear combination for each observation. Thus each principal component (PC) can be expressed (Campbell 1998) as $u_n = a_1 x_1 + a_2 x_2 + a_3 x_3 ... a_p x_p$ where u is the value or score of the new variable or principal component for the nth observation; $x_1, x_2, x_3 .... x_p$ are the values of the original p variables (dimension p); and $a_1, a_2, a_3, ... a_p$ are the assigned weights for each variable.

The principal component analysis has been employed for across studies, for instance, to analyse 60 epidemics of papaya ring spot virus, each characterised by 10 variables. Mora-Aguillera et al. (1996) initially placed variables into two groups to reduce collinearity among variables. Each group of variables was then evaluated by PCA to identify principal components (PCs) that accounted for a small amount of variance and to eliminate variables using minimum principal components (following Hawkins and Fatti 1984). With a successive series of PCA and inspection of the biplot display, three variables were detected, viz. standardised areas under disease progress curve, the c-value of the Weibull growth function; and X_0 (i.e. $t(y_0)$) for the time in days from transplanting in the field until the first symptoms. These variables were retained to provide an overall description of the epidemics, accounting for 83% of the variance of the original variables.

PCA with or without rotation of the correlation matrix may be used instead of factor analysis (FA). Navas-Cortés et al. (1998) calculated a principal component with an SAS FACTOR procedure. After the initial factor extraction, an orthogonal varimax rotation was used to estimate the factor loadings to compare and classify the effect of control measures on cultivars (p. 171). Whereas PCA derives principal components from the observed variables, Factor Analysis transforms the underlying factors to observed variables (Mardia et al. 1979). The factor loadings b_{ij} represent the correlation of the variable i with the factor j and a residual component E that is specific to the variable i and is not related to any of the other factors. Variables x_n are linear functions of hypothetical functions called factors f_n. With factor analysis groupings of elements of the disease progress curves for their inter-relationships were obtained in various factors, thus furnishing broader information (Tables 5.1 and 5.2). For comparisons of structural relationships among pathosystems, path coefficient analysis may be an appropriate statistical technique for the evaluation across multi-factorial experiments in comparative epidemiology (Jörg 1987; Hau and Kranz 1990; Khoury and Kranz 1994).

ANOVA and MANOVA are suitable for a statistical comparison of parameters in growth functions, e.g. the regression coefficient b, rate and shape parameters and AUDPC-values. Prerequisites and interpretation of their results of ANOVA are discussed in detail by Gilligan (1983), Madden (1986), Campbell and Madden (1990) and Petraitis (1998). A special case of ANOVA is the repeated-measures analysis of variance described by Campbell and Madden (1990) with the Wilks lambda function as test statistics. Campbell (1998) discussed details and Littell et al. (1991) critically review this method. It is an alternative procedure to ANOVA if disease progress, due to interaction effects between treatments and time, cannot be analysed by growth functions or AUDPC. Repeated-measure designs are similar to split-block design. The diseased plants could be the plot or group of plants on which disease is assessed at each time for changes in disease intensity. The pitfalls of this analysis is that the F-values for time and the time-by-cultivar interaction will be too large in the case of a high autocorrelation, reflecting a higher degree of significance than is warranted.

Another statistical evaluation of comparative results is possible with Chi^2 test (p. 30) of contingency in row and column (r×c) tables (r number of samples with c characteristics) for significance of independence (or inhomogeneity), to test the goodness-of-fit of nominal scaled variables and homogeneity (whether samples belong to the same population). If coefficients of association derived from r×c tables of different size are to be compared, the normalised coefficient of contingency would be preferable if $0 \leq C = \sqrt{m \times chi^2 / [(m-1)(chi^2 + N)]} \leq 1$ (Köhler et al. 1996) with N = sample size, m = the smallest values of r and c in the rows and columns of the contingency table. With this normalised version various contingency coefficients can be compared. C-values are comparable to the correlation coefficients r or R, by Pearson and Spearman, respectively.

Madden (1986) and Madden and Campbell (1990) treat statistical analysis of disease progress curves and their comparison in great detail. For a comparison of disease progress in time and space across epidemics, confidence intervals (CI) can compare directly the estimated parameter values from at least two epidemics, provided the same growth function and their derivatives (e.g. rates) were used for each epidemic. Rates estimated from a logistic function cannot be compared directly with the one from a Gompertz function or Bertalanffy-Richards. A weighted absolute mean rate has to be computed first. All data points that fall inside the confidence interval then would be equal and those outside the interval are different. Campbell and Madden (1990, in their Sect. 8.5.1.1) present an example of how to compute a CI. They used data of disease progress (Gompertz function) of potato late blight over 37 days (seven assessments) on two cultivars (Campbell and Madden 1990, their Table 8.2). From this, the following equations 8.30 and 8.31 (in their text) were established

$$(\Theta_1-\Theta_2)\pm t[p/2; n_1+n_2-(2p)]s[d]$$

$$s[d]=(s^2[\Theta_1]+s^2[\Theta_2])^{1/}$$

where Θ_1 and Θ_2 are estimated parameters (r in this case), p is the number of parameters in each function, n_1, n_2 the number of observations for both epidemics in the example, t is the critical value from a t-table (one-sided) with significance level P, and $n_1+n_2-(2p)$ degrees of freedom and s[d] is the standard error of the difference. Finally, the residuals of regression lines should be tested for autocorrelation. With these equations Campbell and Madden (1990) had infection rates r=0.131 of cv. Sebago with s=0.004 and of cv. Katadhin r=0.214 with s=0.008. From Eq. 8.31 s(d)=0.009 and the confidence interval from Eq. 8.30 results for 95% in CI=(0.131–0.214)±(2.0)(0.009) with t[0.025; 28+28–(2)(2)]=2.0 with 52 df (7 assessment times and 4 replications=28). The calculated interval 0.083±0.02 does not include zero, therefore the rate parameter values are different at α=0.05. For further details, see Campbell and Madden (1990) and Neter et al. (1985).

More statistical methods may be explored for use in comparative epidemiology from other disciplines of science, like taxonomy, ecology and phylogenetics. Gittleman and Luh (1992) list procedures which might be useful in comparative epidemiology as well, e.g. maximum likelihood procedures, nested analysis of covariance, independent contrast comparison and the auto-regression approach. The independent contrast comparison is essentially a procedure to calculate comparisons between pairs of taxa at each bifurcation of a known phylogeny (Harvey and Pagel 1985, cited by Gittleman and Luh 1992). Contrasts are explicitly model-dependent, which provide a process whereby parameter (or trait) variation at successive nodes (e.g. of phylogenetic branching) that are independent of one another can be found. Generally, independent contrast methods use all of the variation in a trait for comparative results. Applications of this method include comparisons of longevity and fecundity in parasitoid hymenoptera, blood parameters among mammals and relative testis size and sperm competition in birds (Gittlemen and Luh 1992). In epidemiology, this method could be applied when taxonomic relationships among pathogens are an important issue in a comparison, e.g. homology or analogy of structures and mechanisms for infection, sporulation (inoculum formation), survival, reactions of parameters of the disease progress curve to ecological factors.

The auto-regression approach in phylogeny divides parameter values into a phylogenetic (inherited) component and a specific component due to independent (adaptive, environmentally influenced) evolution. The method was proposed by Cheverud et al. (1985) for comparative analysis. A phylogenetic matrix reflects phenotypic similarities anticipated at defined phylogenetic distances. The qualities of an individual species are calculated as weighted averages of the qualities of all other species. Close relatives share more recent ancestors and thus are more similar. The greatest weights are assigned to them, whilst distant relatives have low weights. A variable weighting index, like the maximum likelihood procedure, provides greater flexibility to deal with both conservative and plastic phylogenetic

traits. In phylogeny, for instance, with this method, body size dimorphism in primates, brain size and life history patterns across mammalian carnivores, ornament
dimorphism and mating systems in birds were compared. Moran I or other descriptive statistics may be less efficient for such a comparison. In epidemiology,
autocorrelation is used in the analysis of individual experiments (Neter et al. 1985;
Madden 1986; Campbell and Madden 1990). Campbell and Benson (1994a) describe its application for spatial autocorrelation analysis of soilborne diseases.
Contrary to phylogeny, the latter authors recommend the Moran I coefficient to
elucidate spatial processes during root disease epidemics. Autocorrelation could
be a valuable criterion for comparisons of dynamic processes across pathosystems.

Apart from the commonly used parametric tests, non-parametric tests may also
be of use in comparative epidemiology, for instance, the Friedman Kruskal-Wallis
tests for unifactorial analysis of variance and rank (the latter commonly used for
independent samples). For multiple comparisons of paired samples, the Wilcoxon-
Wilcox and Nemenyi tests are more appropriate.

3.3.2.2 The Equivalence of Results

Quantitative results of across studies often look qualitatively similar, but are not
significant when submitted to conventional statistical tests. However, they still
may be equivalent. This dilemma easily arises with features and parameters from
pathosystems and epidemics observed over years or sites, e.g. time of onset of
epidemics $t(y_0)$, of primary intensity y_0, peaks (y_{max}) shift and vary and so do the
shapes of disease progress curves and gradients. For instance, a high probability of
recurrence of peaks in a certain growth stage of host plants across data sets or
pathosystems makes it a plausible fact, while disease intensity and the date at
which y_{max} is reached can vary in each individual study. Standard statistical tests
are then either not applicable, or the results, although apparently similar, are not
consistently significant with common test parameters.

For comparative epidemiology, it might be important not to lose statistically
non-significant, but recurrent information. Various options are available to preserve such data sets (after another critical evaluation) for across studies. Non-
significant parameter values, for instance, correlation coefficient r, may be included in a meta-analysis (Hunter and Schmidt 1990). Frequency analyses of
events that occurred, rank correlation coefficients and tabulations may help to get
around the hard significance tests and arrive at plausible and justified conclusions
on a phenomenon. Tabulation, for instance, ranking disease severity of cultivars
tested for resistance against pathotypes was proposed by Van der Plank (1968;
Vanderplank 1984) to identify whether cultivars have race-specific (vertical) or
race-non-specific (horizontal) resistance and the ranking of the latter. The Spearman rank correlation coefficient is a procedure to compare data ordinally scaled
and increasing monotonously (up or down) without a linear relationship. Similarity and matching coefficients are being used without any test for significance. A
more recent approach to deal with the problem of not being significant at a 5%
level of results is equivalence analysis.

Equivalence analysis, though still debated, is already widely used in medical
and pharmaceutical research mainly for proof of bioequivalence of the therapeutic

equivalence of medical treatments (Wellek 1994; Elze and Blume 1999). The US
Food and Drug Administration (FDA) has accepted average bioequivalence and
drafted a guideline for the development of population and individual bioequivalence, which has been commented on by Röhmel (1999).

In order to comply with scientific standards, without the loss of relevant information, statistical equivalence tests for equal means are proposed to evaluate
whether results for which the null hypothesis the commonly accepted values in
significance tests (i.e. $\alpha=0.05$ etc. for the error H_0) fails, still have a functional
equivalence. When, in a comparison across studies, the α-risk for a hypothesis
should be avoided and not rejected unnecessarily, an acceptably higher value can
be chosen, e.g. 10% for α (Wellek 1994). The value for error H_1 (β), which accepts the null hypothesis, though it is actually wrong, then becomes smaller.
Equivalence analysis thus is a softer test sensu Zadoks (1978) to establish identity
of two or more means. For this reason it could assist the selection and use of data
for meta-analysis. For an in-depth reading on equivalence analysis and its mathematical methods and implications see Westlake (1976), Wellek (1994), Erickson
and McDonald (1995), Elze and Blume (1999) and Pigeot (1999).

Cases in plant pathology for which equivalence tests would be adequate if the
usual null hypothesis is not suitable, include the following which have relevance
for epidemiology (adapted from Garrett 1997):

1. Is a host or pathogen modified by mutation, genetic engineering, or breeding
   still equivalent, i.e. at least as efficient, to the non-modified original organism,
   or standard, with respect to some or all relevant characteristics except for the
   intended modification?
2. Is disease intensity the same or (for one-sided equivalence tests) at least as low
   for a cheaper or safer management strategy, e.g. IPM, new fungicide, biocon-
   trol?
3. Does a cultivar with a potential for higher yield have the same or at least no
   less resistance than the standard?
4. In general, does a new approach with certain advantages at least perform as
   well as a standard best approach?

Still the best-known, understood and widely applied criteria in equivalence tests
are confidence intervals (Sect. 3.3.2.1), which usually have symmetric intervals to
the mean. Garrett (1997) exemplifies the application of the confidence interval in
equivalence analysis in Fig. 3.4. An a priori-tolerance level for type I error (H_0),
α, is set arbitrarily wide, e.g. >5% as reasonably and biologically justified. In
contrast, the level for the type II error (H_1), β, could be more limited, e.g. $\beta < 5\%$.
Whilst statistical computer packages calculate a P-value for standard hypothesis
tests, e.g. $\alpha=0.05$ (where $p \geq \alpha \rightarrow H_0$, $p < \alpha \rightarrow H_1$), Garrett (1997), for her equivalence test, suggests a P-value for the upper tolerance (see below). As the data were
assumed to be nearly normally distributed, a t-distribution and a standard null hypothesis of equal means can be used to calculate a confidence interval to determine whether it includes zero, otherwise the null hypothesis is to be rejected. With
different means in a stud‚y their confidence intervals are tested to see whether any
of them lie outside the predetermined tolerance.

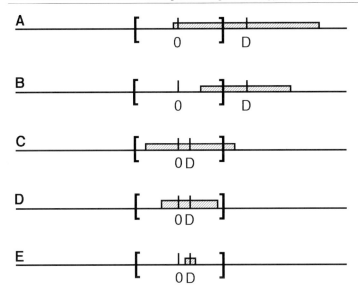

Fig. 3.4A–E. Examples of scenarios testing hypotheses. *Brackets* indicate the equivalence tolerance around 0; the *shaded region* indicates a confidence interval around the observed differences in treatment means (*D*). **A** The confidence interval includes zero, the observed difference falls outside the tolerance; neither the null hypothesis of equal means nor the null hypothesis of a non-trivial difference in means is rejected. **B** The observed difference falls outside the tolerance, the null hypothesis of equal means is rejected as zero is not included because the confidence interval is narrower, while the null hypothesis of a non-trivial difference in means is not. **C** The observed difference falls within the tolerance, neither null hypothesis is rejected. In case **D** the observed difference falls within the tolerance; the null hypothesis of equal means is not rejected, while the null hypothesis of a non-trivial difference in means is rejected. **E** The observed difference falls within the tolerance, both the null hypothesis of equal means and the null hypothesis of a non-trivial difference in means are rejected. (Garrett 1997, with permission)

In Garrett's example (Fig. 3.4), the confidence interval extends above the upper tolerance, so the equivalence null hypothesis is not rejected. The *P*-value for the upper tolerance was calculated in Garrett's example as follows: an "... experiment is performed to compare a new management technique with a standard technique for controlling powdery mildew of roses. The experimenter, planning to measure percent infection at the end of the experiment, might determine a priori that the difference of 5 in percent infection is insignificant from a practical standpoint. If the mean percent infection of 10 indicates that the new technique is 12 and of the standard technique 9, the observed differences between means is 3. Using a standard null hypothesis of equal means, the experimenter would construct a confidence interval around this estimate and determine whether it includes zero. Suppose the data are approximately normally distributed, so that an interval based on a t-distribution can be used. If the pooled standard deviation is 6.1, then a 95% confidence interval around the estimated difference is $3\pm2.1\times6.1\times0.45$, or 3 ± 5.8, in which 2.1 is the critical t value for 18 degrees of freedom (df) and $0.45=\sqrt{1/10+1/10}$. This interval includes zero, so a null hypothesis of equal means is not rejected. To test the equivalence null hypothesis of different means, the researcher determines whether any of the confidence interval lies outside the predetermined tolerance, –5 to +5. The confidence interval extends above the upper tolerance 5, so the equivalence null hypothesis is not rejected either. For this experiment, a larger sample size or lower variance would be needed to determine conclusively whether the mean effects of the techniques are different or equivalent at the 95% confidence interval" (Garrett 1997). The P-value for the upper tolerance of the equivalence test example can be calculated from the difference observed between the means, 3, and the upper tolerance, 5, which is 2. This difference is scaled by a standard error $2/(6.1 \times 0.45)=0.73$. For a t-distribution with 18 df, 0.73 is at percentile 0.76, giving a P-value of $1-0.76=0.24$. Percentiles can be found in statistical programmes.

Figure 3.4 illustrates a series of possible outcomes A–E from analyses of equivalence. Garrett (1997) concludes "...when a hypothesis test of whether treatments are equivalent is desired, equivalence tests offer a more appropriate framework than the standard null hypothesis. By using a null hypothesis of treatment differences, they place the burden of proof on the experimenter to demonstrate that the treatments are equivalent". However, when comparisons are done with several levels of continuous treatments or measuremants, a regression analysis would be more appropriate. Further types of equivalence test are referred to in Garrett's paper.

3.3.2.3 Classification by Similarities

Comparisons of epidemics and their characteristics may lead to classification. For practical purposes, it would suffice to only compare components of epidemics which, for instance, have consistently shown important effects on the behaviour of an epidemic in sensitivity tests (p. 30).Most simply, classes may just be named after well-described epidemics, like the Heines VII epidemic of wheat stem rust (Zadoks 1961), or the northern Japan type of rice blast epidemic (Kato 1974). Classes may also be characterised by growth functions or the shapes of disease progress curves typical for a pathosystem, e.g. monomolecular or logistic, or uni- or bilateral, bimodal etc. respectively. Types of epidemics and their models could be stored as yardsticks in computers for a rapid comparison with any new epidemics obtained.

Principal component analysis (PCA), factor analysis (FA) and in some also the ANOVA, are appropriate for comparisons of structural elements (Tables 5.1 and 5.2) and intrinsic features such as threshold, limits, recurrences etc. Cluster analyses (CA) and discriminant analysis (DA) are specific statistical methods for classification . Cluster analysis forms groups (clusters) of elements from a given data set (quantitative criteria, qualitative attributes, e.g. from the disease progress curve) which are more similar to each other than to those remaining outside a first, second, and so on cluster (Kranz 1968c; Hau and Kranz 1990). Many of the statistical packages offer appropriate, more or less robust routines for cluster analyses (Schlösser et al. 1999) and classification. Cluster analyses are hierarchic, i.e. the process of selection for a cluster starts from the variable with the highest weight in the data set. The elements of any new cluster extracted from the data set are clearly distinct from the subsequent ones, sometimes with decreasing linkage distance of their variants. As a measure of linkage distance the squared Euclidian distance may be used. The final number of clusters can be obtained by calculating the pseudo F and pseudo t-statistics. By means of the WARD method (SAS Statistical Package 1988), its pseudo-f-value (ratio of between-cluster sum of squares and within-cluster sum of squares) may be calculated to make a ranking possible.

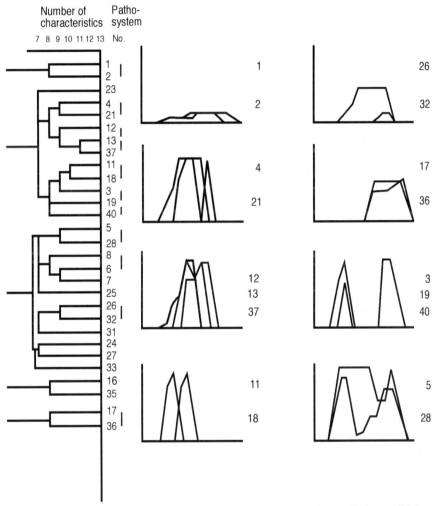

Fig. 3.5. Dendrogrammes after cluster analysis from 40 epidemics studied on wild host plants. For each of them, 13 criteria were assessed over 2 years. The graphs of epidemics are shown (*numbers*). Epidemics with a high resemblance in disease progress curves are linked by *bars* (*vertical lines*). (Kranz and Lörincz 1970; Hau and Kranz 1990)

Any procedures of cluster analysis for each variant require a number of criteria as descriptive attributes, nominal, ordinal or metric parameters and similarity and matching coefficients (Sokal and Sneath 1963; Sneath and Sokal 1973; Clifford and Williams 1976) to identify the degree of similarity. These are displayed graphically either as a dendrogramme (Figs. 3.5 and 7.5), cluster tree, as a projection of cluster on a two-dimensional plane of two components, or in space with three principal components (three-dimensional plot). Procedures may give different results, or differ in some aspects. For instance, the procedure MCEF (Kranz and Lörincz 1970) gave a slightly better fit of groupings than CLUS and permitted a fully automatic comparison. Weighting of elements in these experiments did not always improve the results. A first example of a dendrogramme after cluster analysis in epidemiology is the grouping (Fig. 3.5) of epidemics from field experiments for a 2-year period (Kranz and Lörincz 1970). Figure 3.5 emanated from a 2-year study of 59 pathosystems on wild host plants (p. 19). All disease progress curves seemed to have a morphology of their own. After the epidemics were characterised by means of 13 curve elements (see Table 5.1), 40 of these pathosystems could be grouped with cluster analysis into 11 and 16 clusters for the humid and the dry year, each comprised 2–9 pathosystems. Similar clusters were obtained with the programme MCEF (Fig. 3.5). However, only 18 out of 40 pathosystems were assigned in each year with their curve model to the same cluster. Out of 18 of these pathosystems grouped in the first 5 clusters in the wetter year, 13 appeared again in the same first 5 clusters of the drier year and 10 more in at least adjacent clusters. Shapes of disease progress curves may differ slightly within the same cluster as this was not the only criterion for the grouping. However, it is apparent from Table 5.6 that a pathosystem may have varying DPCs, e.g. one in a dry and another one in a wet year, although the same disease may appear in a particular range or families of clusters (Kranz 1974a). For a comparison of epidemics, it could be useful to know the range of variation, whether changes in patterns are consistent or not and what pattern(s) a disease can attain under varying conditions. Each disease probably has only a limited scope of variation in its disease progress curves. Such knowledge may enhance our capacity to predict future disease progress for better disease management.

A promising and useful statistical technique for classification is DA, the discriminant analysis (Mardia et al. 1979; Hau and Kranz 1990). In contrast to cluster analysis, discriminant analysis have a priori focal points for known or plausibly assumed standards, characteristic types, localities etc., to which incoming entries can be assigned (e.g. pathosystems, epidemics, pathotypes). For discriminant analysis methods a number of variables have to be measured for each entity in the data set before DA separates them into groups with a higher degree of in-group similarity (Fig. 3.6). The pathogen species included in the DA of Fig. 3.6 were grouped according to their vegetation type by means of Venn-diagrammes (Fig. 6.8), where more details on the data are given.

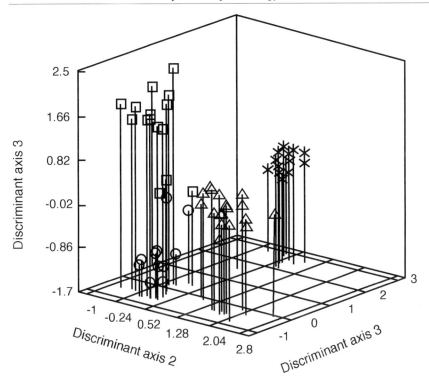

Fig. 3.6. Multiple discriminant analysis to differentiate the pathogen mycoflora of four vegetation types: ground vegetation of forests (*stars*) and forest clearings (*circles*), as well as dry pasture (*triangles*) and meadows (*squares*). Maximum number of pathogen species included were 40. (After Kranz and Knapp 1973; Hau and Kranz 1990)

Beyond procedures of statistical classification, it is tempting to think of a checklist with defined criteria as qualitative and quantitative descriptors (Kranz 1988b). Such a proposed checklist could help to define types or classes of similar epidemics to which any other pathosystem could be assigned. A tentative set of criteria as descriptors is proposed in Table 3.11 (see also Tables 3.1, 3.2, 4.3 and 4.4).

Table 3.11. Some suggested epidemiological criteria as descriptors to classify patho-systems and their epidemics

Component	Criteria as	
	Descriptive term	Measurement term
Life cycle	Monocyclic, polycyclic, polyetic; continuous, discontinuous[a]	
Host		
Range	Narrow, wide, host	Number of host(s)
Resistance type, e.g.	Vertical, horizontal resistance	Delay Δy_o[b] (days);
Ontogenetic resistance	Juvenile, adult plant resistance	y_o, y_{max}, at growth stage of host
Pathogen/disease		
Source of Q	Soil, seed, debris, vector, old lesions, other host(s)	Q[c], Inoculum density (#/mm^2/d) per lesion
Speciation	Formae speciales, pathotypes	Number of pathotypes
Epidemic competence[d]	Fitness, r- or K-strategist, most vulnerable or effective "link" of the infection chain	Range or means of: latent period p, infectious period i, inoculum capacity as #/mm^2/i per lesion, infection rate r, max. disease intensity y_{max}, progeny/parent ratio L, survivability of inoculum (time unit)
Dispersal	Wind-, water-, vector-seedborne	Distance (in m, km)
Essential external factors[e]	Specific, indifferent, narrow, wide for	Temperature °C, leaf wetness h, min, relative humidity % rain mm/time unit, degree days for latent period
Epidemic		
Temporal patterns	Slow rusting, early, late, explosive, mono or bilateral, or bimodal epidemic	Mean Δy_o, mean duration of epidemic (days), y_{max} common at growth stage host, range of AUDPC
Model type	Unilateral, bilateral bimodelar, wave-like	Fitting growth functions[f]
Spatial patterns	e.g. Focal spread steep or flat gradient, isopath	
Model type		Exponential or power law for gradients

[a]An epidemic is continuous (Gäumann 1951) when the inoculum comes from the same host(s), i.e. the infection chain is not interrupted (e.g. cereal powdery mildew with spring and winter crops). Discontinuous infection chains either have a saprophytic phase, resting spores, or pathogens with heterogeneous life cycles, i.e. with obligately alternating hosts, like macrocyclic rusts.

[b]Delay Δt in days between emergence of crop and onset $t_{(40)}$ of disease.

[c]Q stands for overseasoning inoculum in soil, seeds, debris, old lesions on the host plant, vectors etc. available to cause y_o. Appropriate measurement terms may be established experimentally.

[d]For features (or parameters) of fitness see Table 4.15 and Fig. 4.12, for r- and K-strategists see Fig. 4.6 and for components of the infection chain see Tables 4.4 and 4.5.
[e]See Table 4.5. Important compensating factors may be identified, e.g. longer leaf wetness duration for suboptimal temperature. Also, factors connected with cultural practices (e.g. crop rotation, fertiliser, plant density) may be added as a specific criterion for the patho-system.
[f]See Fig. 5.3 and Table 5.13 and related text.

Similar to suggestions made in Table 3.11 are queries raised by Gilligan (2002) capturing the essential dynamics of epidemics. This may be of importance when deciding how much biological detail should be included in a model to describe adequately the temporal and spatial dynamics of epidemics. Gilligans's reflections focus on the invasion and persistence of pathogens in host populations and on the variability within and between epidemics, which he summarised as:

1. Which biological issues should be included in a model?
   - inoculum dynamics and the dynamics of primary infection
     from resident inoculum or immigration;
   - dynamics of secondary infection between the infected and susceptibles;
   - host dynamics including the rate of susceptible tissue;
   - host responses to infection, including the effects of disease on the supply
     of susceptible tissue;
   - "quenching" due to a change in host susceptibility or the onset
     of unfavourable conditions;
   - vector or antagonist dynamics;
   - genetic structure of the pathogen (and host) population;
   - definition of scale for the susceptible and infected units, ranging from le
     sions, leaves, stems or roots, single plants to whole fields;
2. With which dynamic issues can the model cope?
   - non-linearity (small changes that have big effects on epidemics);
   - stochasticity (chance variation within and between epidemics);
   - temporal heterogeneity (daily and seasonal changes that affect spread);
   - spatial heterogeneity (local and regional changes that affect spread);
   - more than one scale (from micro- through field to regional scale);
3. What will the model be used for?
   - to predict invasion of a new pathogen or race (pathotype);
   - to predict the consequences of control strategies on invasion and persistence,
     or crop loss;
   - to predict the risk of disease.

These points will resolve around the "fundamental considerations" whether or not the models are deterministic or stochastic, whether they are non-spatial or spatial and on the complexity of the model required for the purpose.

4 Comparative Epidemiology at the Systems Levels Host, Pathogen and Disease

The following four chapters will attempt to give an answer as how to approach comparative epidemiology across pathosystems by experiments and posterior analysis. Chapter 4 is devoted to criteria and procedures for CE at the systems levels "host and pathogen" as components of the higher systems levels disease and epidemics. Criteria referred to in the following chapters are, of course, not only valid for studies across pathosystems, but can equally well be employed in within-studies of individual pathosystems or epidemics. Chapter 4 starts with a review of host criteria relevant to comparative epidemiology for monocycles of diseases (Sect. 4.1), followed by the life cycles of diseases (Sect. 4.2), and the components of the infection cycle or infection chain (Sect. 4.3). In Section 4.4 factors will then be compared which determine the epidemic competence of pathogens. At the level of the pathosystems, temporal and spatial features of epidemics (Chaps. 5 and 6) will be compared, and finally in Chapter 7, some effects epidemics can cause.

4.1 Relevant Host Criteria

Pathosystems consist of a host plant and a pathogen under prevailing environmental conditions. A plant species usually is vulnerable (Robinson 1976) to a limited number of the many thousands of potential parasitic agents. It becomes a host to a potential pathogen which, in a given environment overcomes, during the infection process, its defence mechanisms to form a pathosystem, which causes disease. In a pathosystem the pathogen recognises the host plant species as compatible.

Host plants integrate all external factors from the environment and cultural practices, which, in turn, affect their interaction with pathogens through varying levels of susceptibility. The dynamics and effects which result from such an interaction can be compared with criteria from host growth and development (Sect. 4.1.1) including plant morphology, density and cropping patterns, resistances (Sect. 4.1.2) and related sensitivity of yield production. In addition, the interaction between host growth and disease progress will be reviewed as a topic of comparative epidemiology (Sect. 4.1.3).

4.1.1 Host Growth and Development

During their growth and development, host plants undergo changes which affect their interactions with their pathogens. Descriptive and quantitative criteria for growth and development of host plants are the topic of this section. Growth of plants can be measured as length, size and biomass related to time. Seem (1988) reviewed host growth and development in some detail and defined the following parameters as measurement terms for development:

1. GR as growth rate defining the gain in plant mass W in time t as $GR = dW/dt$;
2. GR^* this is GR for the time interval t_1 and t_2 as $(W_2 - W_1)/(t_2 - t_1)$;
3. RGR, the relative growth rate, describes the increase in plant mass per unit mass per unit time as $1/W^* dW/dt$;
4. RGR^* is the mean relative growth rate between t_1 and t_2 as $RGR^* = (\ln W_2 - \ln W_1)/(t_2 - t_1)$;
5. NAR, the net rate of assimilation describing the increase in plant mass per assimilatory material A per unit of time, $NAR = 1/A\, dW/dt$; with a relative rate as $NAR^* = (W_2 - W_1)/(A_2 - A_1)(\ln A_2 - \ln A_1)/(t_2 - t_1)$;
6. $LAR = A/W$, the mean leaf ratio defined as the mean ratio of assimilatory material per unit plant material present. Its relative function is $LAR^* = 1/2(A_1/W_1 + A_2/W_2)$.

These and similar parameters could be applied as well to measure roots and their growth (Campbell and Madden 1990). The most important parameter is root length, and sometimes root diameter.

The development of host plants manifests itself in their phenology. Host phenology (e.g. Popular 1978; Seem 1988; Campbell and Madden 1990) is a descriptive criterion in the form of coded plant growth stages (Table 4.1). Growth stages are indicators for the physiological and ontogenetic state of a crop. They integrate all factors affecting the plant during its development and relate to the phase-dependent (ontogenetic) variation of cultivar resistance (Fig. 4.1), and their relative contribution to final yield. Hence, the same level of disease intensity in more productive growth stages potentially causes a higher loss than in a stage that contributes less to yield (Fig. 7.1). In cereals (Table 4.1), these stages would be tillering, kernel formation and grain filling. Hence, coded growth stages have to be recorded when assessing disease intensity for proper comparison, for modelling, and for crop loss assessment. For posterior comparisons of crop loss the assessment of disease severity in defined growth stages is a must (see Sect. 7.1). Keys for growth stages have been published for various crops either as tables or diagrams, e.g. in a supplement of the FAO Manual on Crop Loss Assessment (Chiarappa 1981) and also in Campbell and Madden (1990). Populer (1978) refers to methods to measure host plant susceptibility. A number of examples are documented in the literature of changes with host age from juvenile susceptibility to adult plant resistance. In plant populations, like cereals, changes in susceptibility of pure lines of host plants are roughly synchronised. However, it can be different in deciduous trees when refoliation is scattered over a period of time, as in *Hevea brasiliensis* (Populer 1972, 1978). The leaves of the rubber tree are susceptible to *Oidium heveae* after unfolding for about 2 weeks only. This affects the epidemic of *O. heveae* if the proportion of young, susceptible leaves is scattered over a longer spell of time in the plantation during periods of favourable weather (Fig. 4.1). Juvenile susceptibility is also known from other pathosystems, e.g. apple leaves can be infected by *Venturia inaequalis* during a period up to 40 days after unfolding (Analytis 1973). Some cultivars of spring barley tend to develop adult plant resistance to *Erysiphe graminis* (Fig. 4.2). Figure 4.7 shows the contrasting juvenile susceptibility of barley to powdery mildew (as in Fig. 4.2) and juvenile resistance to ear blotch of wheat which then reversed to adult plant resistance and susceptibility, respectively.

Table 4.1. The essential criteria of the decimal code for growth stages of cereals. (Zadoks et al. 1974)

Code	Host development in growth stages
00–09	Germinating
10–11	Growth of seedlings
20–21	Beginning of tillering
30	Beginning of elongation
31	First node at the tiller base
32	Second node, next leaf appearing
38	Last leaf visible, but still rolled up, ear beginning to swell
39	Ligula of last leaf visible
40	Booting (heading) begins
45	Sheaths of last leaf fully developed
50–51	First ears (panicles) appears
52–53	One quarter of ears (panicles) appeared
54–55	Half of ears (panicles) appeared
56–57	Three quarters of ears (panicles) appeared
58–59	All ears (panicles) have emerged
60–61	Beginning of anthesis
68–69	Full flowering
71	Flowering is over, kernel watery ripe
75	Kernels milky ripe
83–87	Kernels soft, but dry
91	Kernels hard
92–100	Plant maturing

Morphology and growth habit of host plant cultivars also can affect the severity of epidemics, like the one of *Pyricularia grisea* on rice, mainly due to a difference in spore catches (Schlösser et al. 1999). Morphological characteristics like broad and narrow, erect and drooping leaves thus can be employed as descriptive terms. Leaf area (cm^2), numbers of leaves, branches etc, or height, length and width of stems (cm), dry weight (g) of plant parts are appropriate measurement terms, from which a variable "morphology" was derived by Jörg (1987; Table 3.5). In a path correlation analysis (Fig. 4.15) this variable had the greatest effect on yield. A rather crude though useful measurement term for biomass, i.e. available host plant tissue, is the leaf area index LAI [1], which relates the area of leaves (plus stems and branches) of a plant to the unit area ground covered by the canopy of a plant.

More accurate and detailed assessments are required in exact field experiments (Sects. 3.2.1.1 and 7.1). Foliage of different size or susceptibility may be divided into classes, newly emerging organs may be tagged to follow their life history (e.g. emergence and disappearance). Measurement of leaf area in field experiments can be facilitated by length and width and multiplied by an established coefficient for different classes of age and size. Parameters suitable as criteria for more advanced crop loss functions and host plant-disease models are the LAD [T] for leaf area duration and HLAD [T], the healthy leaf area duration (Seem 1988), and disease assessment (Waggoner and Berger 1987). For research purposes more exact measurements may be needed using electronic equipment to measure relevant units of the host plant, (e.g. for a foliar disease mean leaf size in cm^2 × mean number of leaves/plant × mean number of plants/m^2=mean leaf surface/m^2).

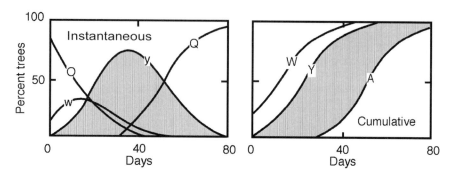

Fig. 4.1. Phenological curves for a clonal rubber tree population about wintering time. The percentages of trees are given for old leaves (Q), of defoliated, wintering trees (W), trees with new growing and susceptible leaves (Y), and trees with new adult leaves (A). The same data are plotted as instantaneous values (*left*), and as cumulative values (*right*). (Populer 1972, 1978)

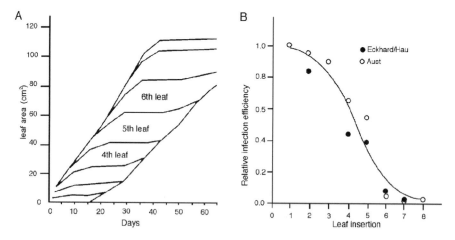

Fig. 4.2. The leaf area and lifetime of the eight insertions (first leaf below) of spring barley as leaves emerge during the season. **A** The proportional leaf area available during the growing season is shown for each leaf insertion, **B** the decrease in their relative susceptibility towards adult plant resistance to powdery mildew with data from Eckhardt/Hau and Aust from the same laboratory. (Hau 1985)

However, biomass of a host plant is not homogeneous in susceptibility (Fig. 4.1) and contribution to yield formation. Older leaves are either more or less resistant, or are already out of the production phase and no longer contribute to the sink "yield". Lateral tillers of wheat were more disease-prone than the main tiller to *Erysiphe graminis*, but less to *Stagonospora nodorum* (Jörg et al. 1987). This may make partitioning of above-ground plant parts necessary, for instance, in leaf insertions (Fig. 4.2) as their ontogenetically determined level of resistance alters the carrying capacity K (Sect. 5.2.4) of a crop available for infection by a pathogen.

The density of a host-plant population also changes as a consequence of their growth and development. Plant density is an important host factor that affects epidemics as the density influences the number and proximity of susceptible hosts, and the crop climate. Measurement terms to compare within crops are the following criteria: plants/unit area of field, the leaf area index LAI [1], and parameter values from modern electronic methods. In a qualitative review, Burdon and Chilvers (1982) compared the various aspects of plant density and their effects on disease incidence. Of the 69 cases reviewed, intensities of 62% of fungal diseases were higher in dense stands, particularly for soilborne and splashborne pathogens. Some diseases like ergot (*Claviceps purpurea*) and many virus diseases are rather favoured by wide distances between plants. Thresh (1983) emphasises this for virus diseases. There are results from field experiments by Hull and Adams (1968) on groundnut rosette virus: In plots sown on the same date, disease incidence was markedly higher with wide-spaced than close stands. In a specific field experiment with leaf rust of wheat (*Puccinia recondita*), Pfleger and Mundt (1998) could not prove any significant effect of the various initial densities in their plots. The reason for this lack of effect was the increased tillering in low seeding densities, which led to a higher density anyway. In mixed populations of wild host plants as metapopulations, the higher disease incidences of about 50 pathosystems were positively correlated with higher plant density (which enhanced humidity), but not always with disease severity (Kranz 1968a). At a communal level, density can be described by the percentage of area (or number of fields) sown to a susceptible crop (Van der Plank's "popularity"), to distances of potential inoculum sources, etc. Metapopulations may also be distinguished in a population of a given crop by classifying subpopulations according to criteria that are relevant for a comparison, e.g. plants uninfected, latently infected, infectious, and no longer infectious ("removed" sensu; Van der Plank 1963). Any other differentiation of populations may be considered as a metapopulation, e.g. flag leaves and ears of cereals for yield loss assessment functions, spatially diverse plant populations, various pathotypes within pathogen populations.

The principal driving force behind the parameters for growth and development is temperature. Criteria for physiological time are also employed like temperature equivalents expressed as "degree days", phyllochrons (phytochrons), or similar functions. Degree days accumulate the effect of temperature over time. The choice of a minimum temperature is important and will be used as the base above which degrees are accumulated, e.g. 50 °F in the USA. This base temperature applies for all host plants and pests to calculate a standardised threshold value of degree days for different stages in each species. As temperature varies from place to place and season to season, values for temperature equivalents have to be calculated with actual degrees measured by means of available software for various routines. A similar measurement term is the phyllochron, i.e. the time elapsing between the appearance of one leaf insertion and the following one.

Seem (1988) stressed that physiological time should be preferred to a chronological time scale such as DAS (i.e. days after seeding). However, like LAI, DAS is a valuable parameter to be used for routine trials. Whenever recorded, DAS ensures a certain comparability of results which is useful for posterior across studies. Obviously, crop climate effects can be modified by agricultural practice like plant densities and doses of nitrogen. Criteria with a physiological time scale, like degree days, phyllochrons and growth stages, measured and recorded in experiments in relation to disease intensities would improve the prospects for posterior comparison of epidemics. These criteria are also better suited for crop-disease models, provided an appropriate temperature equivalence function like the Beta or general beta functions (Hau et al. 1985) is employed (p. 83).

4.1.2 Resistance of Host Plants

Host development changes the phase-dependent (ontogenetic) resistance (Fig. 4.2) within the framework of the genetically determined resistance of a cultivar. Host plant resistance in a given pathosystem can be identified and compared by the measurement terms in Table 4.2.

Different types of host plant resistance for cultivars are known: e.g. complete/partial, dilatory/rate-reducing, seedling/adult plant, monogenic/polygenic, major gene/minor gene, vertical/horizontal (Welz and Kranz 1997). We focus here on the two categories of qualitative or race-specific, and quantitative or race-nonspecific resistance. These types of resistance have epidemiological effects and they are the least ambiguous descriptive terms. These terms are equivalent to ver-

Table 4.2. Parameters to measure whether host plant resistance is lower or higher when tested with standardised inoculum

Parameter	Resistance is ... when	
	Low	High
Disease intensity[a] (both y_0, y_{max})[b]	High	Low
Incubation period	Short	Long
Latent period p	Short	Long
Infection efficiency	High	Low
Disease efficiency	High	Low
Infection rate r	Fast	Slow
Lesion size	Large	Small
Infectious period i	Long	Short
Sporulation intensity	High	Low

[a]Disease intensity may be incidence or severity.
[b]Initial disease intensity y_0, maximum disease intensity y_{max}, and y_t, the disease intensity measured at any time t during the epidemic.

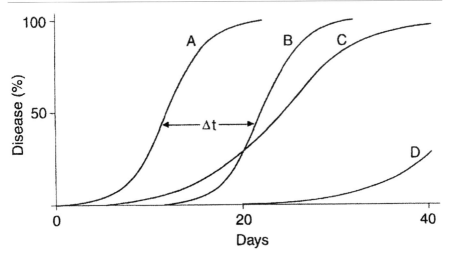

Fig. 4.3. The delay-type resistance (qualitative, race-specific, vertical) and rate-reducing (quantitative, race-nonspecific, horizontal) separately and combined, and their effects on disease progress curves. *Cultivar A* has little rate-reducing and no delaying resistance. *Cultivar B* has the same small rate-reducing resistance as *cv. A*, but has one vertical resistance gene. *Cultivar C* resembles *cv. A* in not having a vertical resistance gene, but has much more rate-reducing resistance. *Cultivar D* combines the delaying (vertical) resistance of *cv. B* and the rate-reducing (horizontal) resistance of *cv. C*. (Vanderplank 1984 modified by Welz and Kranz 1997)

tical (VR) and horizontal (HR) resistance sensu Van der Plank (1963). Both of these types of resistance influence disease progress curves (Fig. 4.3). Vertical resistance is based on gene-for-gene relationships. That is, for a specific gene for virulence or avirulence in the pathogen, there is a specific gene for susceptibility or resistance in the host. Ideally, for cultivars grown under exactly the same conditions with and without qualitative resistance (and the same low level of quantitative resistance), the disease progress curves do not, at least theoretically, differ in shape, but in Δt, the delay caused for the start of an epidemic (disease onset). The curves remain the same on each date of measurement (curves A and B in Fig. 4.3). Quantitative resistance alone, by its rate-reducing effect, will produce curve C in Fig. 4.3. The combination of both, qualitative and quantitative resistance in the same host, results in curve D in Fig. 4.3.

Both qualitative vertical and quantitative horizontal resistance can be differentiated (apart from a visual comparison of disease progress curves) in two ways. First, if several host and pathogen genotypes are plotted in a matrix, a differential interaction between pathotypes and host genotypes occurs in cases of vertical resistance. If the host genotypes have horizontal resistance, then the individual cultivars will retain nearly the same rank against any pathotype. The second method to detect if the resistance is horizontal or vertical is the analysis of variance (ANOVA). A significant F-test for the interaction between host cultivars and pathogen isolates means that the resistance is vertical if the F-test is also significant for host genotypes and pathogen isolates. If their interaction is not significant, then horizontal resistance, QTL (Quantitative Trait-linked Loci), or incomplete vertical resistance genes can be assumed. The effect of these types of resistances on epidemics will be compared in Section 7.3.2.

4.1.3 Interactions Between Host Growth and Epidemic Development of Disease

Populations of hosts and of pathogens are affected by factors from the environment and these factors will affect the shape of disease progress curves and the carrying capacity of hosts for disease in a crop (Sects. 4.1.1 and 4.1.2). The effects of host growth, i.e. plant surface available for infection, will be dealt with later (Sect. 5.2.2.3) in relation to the rate of disease progress and the asymptote of disease. Here, it may suffice to refer to a function that describes the relationship between host growth and disease increase by Campbell and Madden (1990):

$dY/dt=ryY[1-Y/H(t)]$

where dY/dt is the change in diseased area, r is the rate of disease progress, Y is the diseased area in cm^2 as a measure of disease severity, and H is the host area in cm^2 at time t.

 The effects of plant growth on the rate of disease progress was of interest to Van der Plank (1963). He proposed a corrected rate of disease increase to account for an exponential change in susceptible biomass of hosts as $\rho=[1/(t_2-t_1)]\{\ln[y_2/(1-y_2)]-\ln[y_1/(1-y_1)]+\ln[(H_2)-\ln(H_1)]\}$ with H_1 and H_2 for the measurement terms of the biomass of the host at times t_1 and t_2. In experiments with five herbaceous pathosystems (Sect. 5.2.4), Kranz (1975a) employed this function to assess the relationship between leaf area and disease progress (Figs. 5.5 and 5.6). He calculated four growth rates for which the designations used by Campbell (1998) were adopted for the rates. These were: r_{la} for the change of leaf area, r_{dls} change of diseased leaf area, r_{ds} change of disease severity, and r_{cg} the rate of disease increase or decrease corrected for host growth. We will return to this study in Section 5.2.2.3.

 More attempts were made after Van der Plank to evaluate the effect of host growth on disease progress. The reader is referred to Kushalappa and Ludwig (1982), Rouse (1983), Waggoner (1986), Jeger (1987) and Seem (1988). They all include carrying capacity K in the growth functions. The functions by Rouse, Kushalappa and Ludwig, and Jeger analyse the effect of host growth on disease rates. The interactions of changing host, environments, and rates are critically discussed by Waggoner (1986). For across studies, the most convenient and appropriate function may be chosen from the menus presented by Waggoner (1986) and Seem (1988).

4.2 Life Cycles of Pathogens

Most plant diseases are of an infectious nature and caused by pathogens (viroids, viruses, phytoplasmas, rickettsia, bacteria, and fungi). There are, however, also abiotic diseases or disorders caused by nutrient deficiencies or excesses, soil disorders, pollutions etc., which may also exhibit temporal and spatial dynamics. Pathogens have life cycles with fixed sequences of events to survive from one overseasoning to the next (Fig. 4.4A–D). The survival of the pathogens is determined by the vegetation period of the host plant or, with perennial hosts, by periods of favourable weather conditions. During overwintering of pathogens and diseases, death and birth rates are greatly influenced. However, though an essential phase of an epidemic, or between epidemics, overwintering (or overseasoning) has found little attention in comparative epidemiology. Survival of inoculum Q and, consequently, primary infections y_t or the start t (y_o) of the epidemic is determined in the following season. Sexual recombination often takes place during this phase in ascomycotina and basidiomycotina with effects on the pathogen's variability. If there is a saprophytic stage, this may alter the pathotype frequencies reached at the end of the preceding season, with an impact on the longevity of cultivar resistance, or the efficacy of a risk fungicide. Hence, the overseasoning of pathogens and diseases deserves more attention in epidemiological research.

Across pathosystems, life cycles can be compared in descriptive terms by the sequence of biological components, which includes overseasoning, infection, latency, lesion formation, infectious period, and dispersal. Gäumann (1951) termed this invariable sequence the "infection chain". From these components, measurement terms can be derived as criteria to compare elements of the infection cycle (Sect. 4.3).

Life cycles among diseases caused by pathogens can either be mono- or polycyclic. In monocyclic diseases, a life cycle practically corresponds to a single infection cycle. It starts with infections during limited periods at the start of host growth. There is no secondary infection in the same season and crop (Fig. 4.4A). Polycyclic diseases just have several infection cycles to buildup epidemics from inoculum produced in the crop (Fig. 4.4B). If the infection chain is ruptured by adverse factors or control measures, no daughter lesions could emanate from existing lesions and the epidemic will be impeded, if not stopped. For polycyclic diseases, the speed of epidemics depends on the actual values of the monocyclic parameters of their various infection cycles, influenced by meteorological conditions and host susceptibility. Life cycle and dispersal mechanisms of soilborne pathogens differ in some aspects (Table 4.4).

The structures of life cycles of pathogens have a bearing on the course of their epidemics, and disease control. For instance, the epidemics of monocyclic diseases are mainly determined by the primary disease intensity y_o, and they tend to follow the monomolecular function. The disease progress curves of polycyclic diseases with several infection cycles are described by the functions of logistic, Gompertz, or Bertalanffy-Richards etc. (see Table 5.13). Sexual stages (Fig. 4.4B,C) recombine genetic characteristics among fungal pathogens which may change pathogenicity. In rust diseases (Uredinales), as with stem rust of cereals (Fig. 4.4C), this may imply an obligate alternation of hosts. In such a case, the

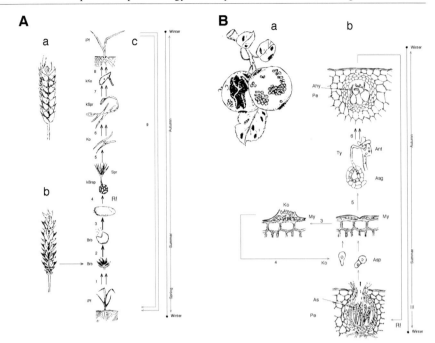

Fig. 4.4. Four types of life cycles of diseases **A** Bunt of wheat: *Brb* smutted grain, *iPF* infected plant, *kBrsp* germinating teliospores, *kKo* germinating conidia, *Ko* conidia, *koSpr* mating sporidia, *kSpr* germinating sporidia, *Spr* sporidia, *R!* meiosis. **B** Apple scab: *Ahy* ascogenic hyphae, *Ant* antheridia, *As* ascus, *Asg* ascogon, *Asp* ascospore, *Ko* conidia, *My* mycelium, *Pe* perithecium, *R!* meiosis, *Ty* trichogyne. **C** Stem rust of wheat: *Ae* aecidia, *Ba* basidia, *Basp* basidiospore, *kAesp* germinating aecidiospore, *kUsp* germinating urediospore, *My* mycelium, *Py* spermogonia, *Pasp* spermatia, *R!* meiosis, *Shy* receptive hyphae, *Tesp* teliospore, *Usp* urediospore. **D** Potato leaf roll virus disease: *Al* alienicola aphids, *iKa* infected potato tuber, *gPf* healthy plant, *kPf* virus diseased plant, *Mi* migrant aphids. (Adapted from Börner 1990)

macrocyclic, heteroecious rusts cause two different epidemics: a monocyclic disease on the alternate host (e.g. barberry), and a polycyclic one on the principle host (e.g. wheat). These macrocyclic rusts produce as many as five spore types in one life cycle. Microcyclic or autoecious rusts, which do not alternate hosts, are monocyclic and they lack pycno-, aecio- and uredinio spores. Finally, vectors are structural elements in epidemics of viruses and other diseases (Fig. 4.4D). Vectors of pathogens, although they have dynamics of their own, are integral parts of the life cycles and epidemics of the diseases they transmit. The dynamics of vectors in vectorborne pathogens parallels the dynamics of wind for windborne pathogens.

Although an infection chain is a principle for pathogens, differences exist as shown in Table 4.3 for three elements across three pathosystems when susceptible growth stages are present. The three pathogens (though having perfect stages) can survive as anamorphs.

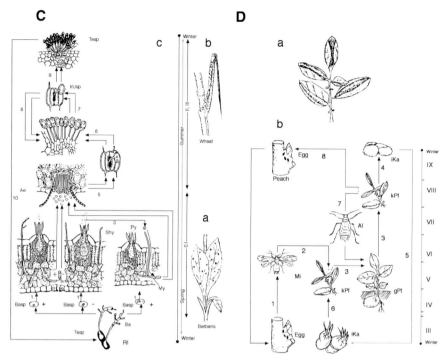

Fig. 4.4. C, D

Table 4.3. Comparison of three elements (or components) of the infection chains in three pathosystems. (Ogawa et al. 1967)

Structural element	*Monilinia laxa* on almond	*Sphaerotheca pannosa* on plum	*Pseudoperonospora humuli* on hop
Infection on	Blossoms, mono-cyclic homogeneous	Fruits, mono- or bicyclic, heterogeneous	Shoots (leaves, stems, stipules) and crown; polycyclic, homogeneous
Sporulation on	Blighted blossoms, twigs, mummies; recurrent	Shoots of roses; recurrent	Shoots (as above); recurrent
Survival on	Blighted blossoms, twigs, mummies and peduncles of previous year	Dormant buds, new shoots of roses	Dormant hop crowns and buds

Among the three fungal diseases in Table 4.3, *S. pannosa* and *P. humuli* both survive in buds, while *Monilinia laxa* survives in diseased plant parts and debris. For the infection of *S. pannosa*, the infection chain is heterogeneous as it may come from an alternative host species, but for *P. humuli* and *M. laxa* it is homogeneous, i.e. with inoculum from the same host species. According to Robinson

(1976), the epidemics of *S. pannosa* begin with alloinfection and the epidemics of *M. laxa* and *P. humuli* start with autoinfection. Despite the same set of homologous elements (Table 4.3), the interaction with host and environment can result in different epidemic behaviour. Life cycles and dispersal mechanisms of soilborne pathogens are presented in Table 4.4. The components may differ among pathogens that temporally invade the soil on plant debris and those that inhabit the soil such as *Armillaria mellea* or *Ganoderma* spp.

Table 4.4. Survival and dispersal mechanisms of soilborne pathogens in annual growing seasons. (Campbell and Benson 1994b)

Life cycle	Initial infection[a]	Dispersal mechanism	Examples
Monocyclic	Direct	Surviving propagules move in soil or plant tissue	*Cylindrocladium crotolariae, Fusarium* spp., *Verticilium* spp., *Macrophomina phaseoli*
Polycyclic	Direct	A. Growing mycelium	*Rhizoctonia solani, Sclerotium rolfsii*
		B. Localised dispersal (zoospores)	*Phytophthora* spp., *Pythium* spp.
		C. Dispersal outside soil by physical forces	*Phytophthora* spp., *Pythium* spp.
Polycyclic	Indirect	Movement of inoculum prior to initial with subsequent dispersal by A,B,C	*Gaeumannomyces graminis, Sclerotinia minor, S. sclerotiorum*

[a]Direct infection primarily occurs through mycelial growth from resting inoculum; indirect infection occurs when intermediate spore type is produced between the resting inoculum and infection (e.g. ascospores from sclerotia in *Sclerotinia* spp.).

Thresh (1983) distinguishes between two categories of monocyclic virus diseases. In one type, the virus spreads exclusively from a fixed amount of inoculum from the soil in a field. These fungusborne viruses remain infective for long periods in resting spores that survive in large numbers and for years between successive plantings of susceptible crops. The second type starts from sources outside a field because vectors either do not thrive in the field or fail to become infectious. Monocyclic virus diseases that originate from alloinfections cause serious diseases even with no or little secondary spread. Vectors of insectborne viruses that cause monocyclic diseases are all winged and are extremely active, at least at certain periods. They invade crops in great numbers when the crop is susceptible. Some monocyclic and polycyclic virus diseases that are spread by various types of vectors are summarised in Table 4.5.

No life cycle exists among diseases caused by non-infectious, abiotic factors (see above). Nevertheless, these diseases or disorders manifest typical symptoms which, like mineral deficiencies, are reproducible in experiments after a process analogous to pathogenesis. They are monocyclic diseases which can become polyetic. A local spread of symptoms occurs when soil conditions deteriorate and nutrients are progressively depleted. The apparent abiotic Fatal Yellowing of oil palms in Brazil seems to be such a case. Temporal (Bergamin Filho et al. 1998)

Table 4.5. Dissemination of virus diseases by different vectors with examples[a] of mono-cyclic and polycyclic diseases on herbaceous crops. (Adapted from Thresh 1983)

Type of vector	Disease cycle	
	Monocyclic	Polycyclic
Contact	Wheat soilborne mosaic (77)	Tobacco mosaic (151)[b], lettuce big vein
Aphid-borne (non-persistent)	Cucumber mosaic (213)	Bean yellow mosaic (40)
Aphid-borne (persistent/semi-persistent)	Beet yellow stunt (207)	Barley yellow dwarf (32)
Beetle-borne	Cowpea chlorotic mottle (49)	Cowpea severe mosaic (209)
Hopper-borne	Maize streak (133)	
	At some sites	At other sites
Thrips-borne	Tomato spotted wilt (39) in tomato	In tobacco
Whitefly-borne	Soybean crinkle	Cotton leaf curl

[a]Examples selected from Tables I and II of Thresh (1983).
[b]Numbers in brackets refer to the CMI/AAB *Description of Plant Viruses*.

and spatial aspects (Laranjeira et al. 1998) of this disease (disorder) have been described. Neither of the exponential, logistic, nor Gompertz functions fitted the temporal disease data well. Spatially, the affected oil palms are aggregated. All the results of spatial analysis were different from what would be expected for the spread of pathogens. On maps of isopathic areas, there was a tendency for affected palms to be found more frequently near river banks and marshes.

Van de Lande and Zadoks (1999), however, believed the same disease in oil palm plantations in Surinam to be infectious. They used geostatistics and gradient analysis to arrive at their conclusions. Geostatistics were found to be affected by size, shape, orientation, scale, and spatial arrangement of the samples. A variogram analysis was used to estimate what is called the semivariance $\gamma(h)$ by calculating the mean of the squared difference between values of pairs of samples for a given lag distance h with an equation from Journal and Huijbregts (1978), $\gamma(h)=\Sigma(Fx_{i+h}-Fx_i)^2/2N_h$, where x_i is the position of a sample unit, x_{i+h} the position of another sample unit at distance h from x_i, F_x the disease incidence measured at location x, N_h the of sample pairs for distance h. The number of sample pairs at each distance was at least 100. The authors fitted three types of variograms, according to Matheron (1997): (1) no correlation between sample points, the variogram is flat (pure nugget effect); (2) the semivariance increases linearly with distance and remains unbounded; (3) the spherical model shows a maximum semivariance at some distance, called the range of spatial dependence. The variograms were compared.

4.3 The Infection Cycle or Chain

4.3.1 The Infection Chain and Factors that Affect it

An infection cycle as the monocycle of biotic diseases, consists of an invariable chain consisting of components, or elements (Table 4.6). For each element, or "link of the chain", there are descriptive and measurement terms (Tables 3.1 and 3.2) which can be used for comparison. They are affected by factors from the host, the environment, and human interference. The actual number of components to be distinguished in a chain is up to the discretion and need of the researcher. For epidemiological studies, the elements in Table 4.6 usually suffice. It is, however, possible to split, for instance, the component of infection into various phases or subcomponents such as: germination, formation of appressoria, penetration, infection efficiency, and infectious period. Components will be examined in Section 4.3.2.

Table 4.6. Essential components of the infection chain as descriptive terms and some of their parameters (measurement terms) as state variables

Component (element)	State variable and dimension
Infection	Infection time [T]
	Infection efficiency [1]
	Infection period [T]
Pathogenesis (incubation)	Incubation period [T]
Lesion manifest (end of incubation)	Disease intensity [1]
	Disease efficiency [1][a]
Latency	Latent period p [T] (onset inoculum formation)
Multiplication (inoculum)	Infectious period i [T]
	Sporulation intensity [1] (spores/mm^2 lesion/day)
	Sporulation capacity [1] (spores/mm^{2}lesion/i)
Dissemination (Inoculum in transit)	Spores caught [1] (numbers/unit/time[b])
	Mycelial growth [1] (length/unit/time)
Survival[c]	y_0 for the next crop [1]
	Duration [T]

[a]Disease efficiency = percent lesions manifest from 100 propagules applied (landed) to (on) area unit of host surface. Infection efficiency refers to percent infections identified either microscopically or from early symptoms that do not develop into lesions (e.g. the yellow streaks of Sigatoka disease of banana). If the density of inoculum is not known, disease intensity (either as disease incidence or severity in %) becomes the criterion.
[b]The actual definition very much depends on the technique used to collect propagules.
[c]Survival, strictly speaking, is part of the life cycle. Short periods of survival (Sect. 4.2.3.6) may, however, occur intermittently in polycyclic diseases.

A variation on the concept of the infection cycle is offered by Bergamin Filho and Amorim (1996) when structural relationships in infection cycles of temperate and tropical pathosystems are compared for existing differences (Fig. 4.5). The possible infection sites on hosts in Fig. 4.5A–C are defined according to Van der Plank (1963) as: healthy sites, latently infected sites (both with no symptoms), lesions (infectious sites), removed lesions (non-infectious sites). Figure 4.5A represents the clockwise infection chain of a temperate pathosystem. The difference seen by Bergamin Filho and Amorim between a temperate and tropical pathosystem is the existence of an anti-clockwise flow of inoculum represented by lesion growth (Fig. 4.5B). Some pathogens with lesion growth are *Alternaria solani* on potato and tomato, *Bipolaris sorokiniana* on wheat, *Exserohilum turcicum* on maize; others with numbers of lesions, e.g. *Microcyclus ulei* on rubber, *Mycosphaerella* spp. on banana, and *Pyricularia grisea* on rice and wheat. The sexual stages of pathogens in the infection chain, e.g. *Microcyclus ulei* and *Mycosphaerella musicola*, are given in Fig. 4.5C.

Bergamin Filho and Amorim (1996) concluded that diseases without an anti-clockwise infection chain are less important in the tropics. Among the cycles of tropical diseases, the inoculum moves in both directions, anti-clockwise and clockwise, the latter for asexual and sexual spores. Spores of perfect stages of fungi play a different, though complementary role, in the epidemic. An obvious example are the various types of spores of macrocyclic rusts. However, in a comparison, a distinction between epidemics that start with spores from sexual and asexual stages may be necessary in some pathosystems.

The specific weight of a criterion in the infection chain (Table 4.4) can differ among pathosystems, epidemics, system levels, or research objectives. The weight may be derived from categories, each category with a different ranking of the components (Rapilly 1991; Table 4.7). Such an order, if consistent, may permit the assignment of weights to criteria with primary, secondary, or tertiary importance. Weighted criteria exert additive, positive, negative, enhancing, or compensatory effects on other criteria in classifying statistics (Sect. 3.4.3), and in models.

For a comparison of monocycles, Zadoks and Schein (1979) proposed the following parameters as criteria: primary or initial disease intensity y_0, latent period p, infectious period i (Table 4.6) and, in addition N as the number of propagules produced (i.e. sporulation intensity) and E the infection efficiency. Depending on whether these parameters of the disease monocycle attain low or high values as weights (Fig. 4.6), a disease can be classified either as an r- or K-strategist (Zadoks and Schein 1979; Fig. 4.6). Pathogens as r-strategists, for instance powdery and downy mildews, react rapidly with quick sporulation and an explosive spread when weather factors become favourable. *Alternaria* spp. are rather K-strategists because they tend to be less sensitive to adverse or favourable conditions, and thus less explosive (Rotem 1994). Very typical K-strategists are a number of pathogens of forest trees, like rust on white pines, *Armillaria mellea*, *Rigodoporus lignosus*, *Ganoderma* spp., *Fomes* spp., and many other soil-inhabiting pathogens.

A

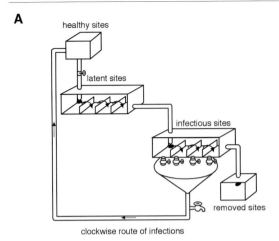

B

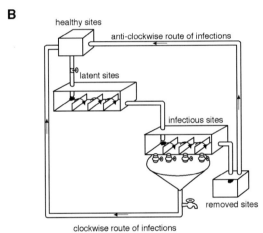

C

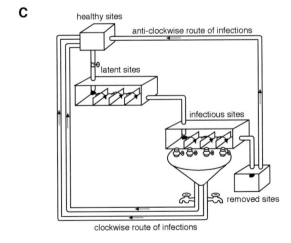

Fig. 4.5A–C. Pictorial comparison of infection cycles of temperate and tropical pathosystems; see text for explanations. (With permission by Bergamin Filho and Amorim 1996)

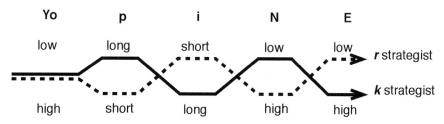

Fig. 4.6. Low or high parameter values of the components (see text) of the infection chain that determines an r- or K-strategist among pathogens. (Zadoks and Schein 1979)

Table 4.7. Importance (in decreasing order) of components from the infection chain in three categories of epidemics. (After Rapilly 1991)

Category	Components	Examples
1	Primary inoculum>infection efficiency>infectious period>latent period	Smuts, downy mildews
2	Latent period>infectious period>infection efficiency>primary inoculum	Rusts, powdery mildews
3	Infectious period>latent period>primary inoculum>infection efficiency	Septoria, etc.[a]

[a]And other leaf diseases, like *Pyricularia grisea, Rhynchosporium secalis* etc.

In simulation experiments Zadoks (1971) confirmed the statement of Van der Plank (1963) that the increase of disease intensity was most rapid with the shortest latent period, the most rapid basic infection rate R, and the longer the infectious period i up to 4 days (see also p. 21). An infectious period beyond 4 days did not enhance the maximum speed of the simulated epidemic. In another simulation, infection efficiency, sporulation capacity, lesion expansion rate (LE), latent period and infectious period were included as parameters for slow rusting of *Puccinia striiformis* on wheat. The epidemic was most affected by the variables latent period, lesion expansion, and sporulation capacity (Luo and Zeng 1995).

Each component of the infection chain is affected by external factors from the disease square (Van der Plank 1963), i.e. host – pathogen – environmental factors – human interference. These factors have criteria of their own. Weather affects diseases and their epidemics as open systems most directly during their life cycles (see Waggoner 1965; Rotem 1978). A number of common primary weather variables and their parameters are measured in plant pathology. For the details on their measurement, we refer to Friesland and Schrödter (1988) and Sutton et al. (1988). The more relevant primary, directly measured, weather variables in plant pathology include:
– temperature (°C),
– relative humidity (%),
– leaf wetness duration (h),
– rain (either amount in mm, intensity mm/h, or number of days with rain),

- radiation (gcal cm^{-2} s^{-1}; PAR[1])
- sunshine duration (h),
- wind (m/sec; km/day; direction),
- soil temperature (°C) and
- soil water saturation (TDR[2]).

All these weather variables with their parameters may be suitable as comparative criteria which affect the components of infection chains. However, derived or secondary variables as parameters (Table 5.16) will be more telling as they more appropriately relate weather factors to components of disease and epidemics. Their application may ease comparison of the otherwise complex relationships among these variables when defined according to pathosystem and research objective. Temperature relationships, like degree days, determine the development of hosts, pathogens and the epidemic. To model temperature equivalents (Schrödter 1965), both the BETE (Analytis 1977) and the generalised beta functions (Hau et al. 1985) are useful.

Weather factors affect various phases of the infection chain differently among pathogens (Rotem 1978; Campbell and Madden 1990). For example, some *Alternaria* spp. need a light induction period for sporulation, others can sporulate without it. Some pathogens require a different leaf wetness duration with or without an interrupting dry period (Rotem 1994). Reuveni and Rotem (1973) investigated, in growth chambers, the effect of *Leveillula taurica* on tomatoes and pepper (i.e. two pathosystems) as affected by humidity. Powdery mildew on peppers developed better at high than at low daytime humidities. Shedding of infected leaves, however, was more pronounced at low than at high daytime temperature. On tomatoes, the epidemics developed better at low than at high daytime humidity. Xu and Butt (1993) compared conditions required to forecast several apple diseases. Among the factors, rainfall and leaf wetness were important for *Venturia inaequalis* and *Nectria galligena*, but not for *Podosphaera leucotricha*. The conidial concentration of *Erysiphe graminis* over barley fields could not be related to a single weather factor (Dutzmann 1985).

Hence, the analysis and comparison of the effects weather variables have on epidemic features need a good knowledge of the biologically determined requirements of a pathogen or pathosystem. In Table 4.8, the requirements of favourable weather conditions vary in the phases of the monocycles of *Alternaria solani* on tomato as used in the first epidemiological simulator EPIDEM (Waggoner and Horsfall 1969).

In an instructive, qualitative review, Hau and de Vallavieile-Pope (1998) demonstrated that not only the cardinal temperatures for infection differed among four windborne pathogens of cereals in four parameters of the monocyclic diseases (Table 4.9). Major factors that affect infection of barley powdery mildew are temperature, high humidity and leaf wetness. *Erysiphe graminis* and *Puccinia recondita* are active in a wider range of temperatures than *P. striiformis*, which is adapted to cooler temperatures. *P. graminis* is favoured by higher temperatures. Although *E. graminis* germinates over a wide range of relative humidity it reaches

---

[1] PAR denotes Photosynthetically Active Radiation.
[2] TDR denotes Time-Domain-Reflectometry.

a maximum at 97–100% relative humidity, but the role of liquid water for powdery mildew of wheat is not yet quite clear (Merchán and Kranz 1986). The infection is favoured after a minimum duration of 2–4 h for *P. graminis*, and 4–6 h for *P. recondita* and *P. striiformis*. However, at suboptimal temperatures, the minimum leaf wetness duration may have to increase up to 16 h as a compensative effect (Sect. 4.4.2). The interruption of the wetness period can impede or stop the infection in some pathogens, or favour it in others, like *Stemphylium lycopersici* (Rotem et al. 1978). The mean values for optimal conditions in Table 4.9 are considered as indications for slow infection cycles (*P. graminis*) and a fast one for *E. graminis* while the other two rust pathogens are classified as having intermediate infection speeds (Hau and de Vallavieille-Pope 1998).

Table 4.8. Weather and host factors that affect the phases of the infection chain of *Alternaria solani* on tomato as defined for the simulator EPIDEM. (After Waggoner and Horsfall 1969)

Weather conditions favourable in phase of the infection chain							
Conidio-phores	→Conidia	→Dispersal	→Lan-ding	→Germi-nating	→Infection	→Latency	→Lesion growth:
Day Warm	Night Cool				Cloudy Warm not hot	Warm	Cool
Wet	Wet			Wet	Wet		Wet
		Rain heavy Wind brisk	Slight Low			Fruit Setting Heavy	
	Foliage Sparse	Dense					

Table 4.9. Parameter values for components of the monocycles of *Erysiphe graminis*, *Puccinia striiformis* f. sp. *tritici*, *P. recondita* f. sp. *tritici*, and *P. graminis* f. sp. *tritici* under near-optimal conditions. (Hau and de Vallavieille-Pope 1998[a])

Pathogen	Latent period (days)	Infectious period	Sporulation intensity[b] (days)	Disease efficiency (lesions/spores %)	Cardinal temp.(°C)		
					Min.	Opt.	Max.
E. graminis	5	13	15×10^3	33[c]	2	12–24	<30
P. striiformis	10–14	30	15×10^3	20–30	0	8	18
P. recondita	7–8	20	2×10^3	40	2	15	<30
P. graminis	7–9	26	20×10^3	20–45	>9	24	<29

[a]Compiled from various sources. For references see Hau and de Vallavieille-Pope (1998).
[b]Spores/lesion/day.
[c]Aust (1981) gives 20% on juvenile and about 2% on flag leaves as disease efficiency for the barley cultivar he worked with, e.g. 5 or 50 conidia, respectively, are needed under favourable conditions to produce a mildew colony.

The range of their variability under field conditions is determined by temperature, age of host tissue, and density of infection as major influencing factors. The temperature effect on latent periods follows negative hyperbolic curves (as in Fig. 4.7, see p. 71). For the four cereal diseases in Table 4.9, the longest and shortest latent periods were >60 and 9 days for *P. striiformis*, 22 and 5 days for *P. recondita* and *P. graminis*, and 28 and 4 days for *E. graminis* (Hau and de Valla-vieille-Pope 1998).

Human interference, such as cultural practices, as part of the disease square can have many-fold effects on components of the infection chain. Palti (1981) reviewed cultural practices for disease control (Table 4.10). All practices which affect the rate r also influence the latent period p, the infectious period i, the rate of inoculum buildup R, and the duration of host susceptibility s. Cultural practices which reduce y_0 are related to R and s only. Irrigation and sowing date and sowing practices affect all the parameters of the infection chain in Table 4.10.

Shade, according to Palti (1981), enhanced disease intensity of barley powdery mildew (*Erysiphe graminis* f. sp. *hordei*), coffee leaf rust (*Hemileia vastatrix*) and coffee berry disease (*Colletotrichum coffeanum*). However, shade reduces intensities of other diseases such as coffee leaf spot (*Cercospora coffeicola*). Huber and Watson (1974) compiled a list of diseases that were favoured either by ammonium or nitrate as nitrogen fertilisers.

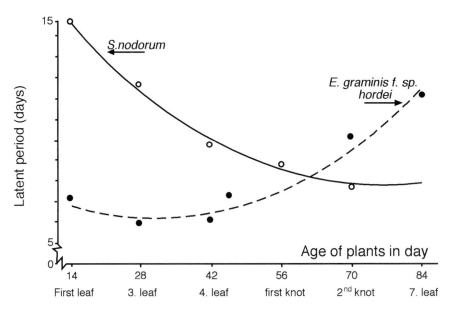

Fig. 4.7. The latent periods of *Stagonospora nodorum* on winter wheat and *Erysiphe graminis* f. sp. *hordei* on spring barley at 20 °C in growth chambers with changing host resistance. (Aust and Hau 1983)

Table 4.10. Cultural practices with effects (+) on primary inoculum y_o, latent period p, infectious period i, apparent infection rate r, rate of inoculum build-up R, and s for the length of susceptibility of the host. (Palti 1981)

Cultural practices which	Affect the parameters					
	y_o	p	i	r	R	s
Sanitation	+				+	+
Crop rotation	+			+	+	+
Tillage	+				+	+
Crop nutrition		+	+	+	+	+
Irrigation	+	+	+	+	+	+
Sowing date	+	+	+	+	+	+
Sowing practice	?	+	+[a]	+	+	+
Harvesting practice	+					
Proximity to source of inoculum	+				+	+

[a]Sowing practice (depth, density) has an effect if temperature and humidity are markedly affected.

4.3.2 Comparison of the Components of the Infection Chain

Components of the infection chain of pathosystems are frequently studied in epidemiology, and these components also are objectives for comparative research. Single components, such as the infection process and factors that affect it, are studied. The entire monocycle may also be studied when forecasts or simulation models are developed. Within an entire monocycle of a pathosystem, sensitivity analysis as an internal comparison is used to identify the component(s) that contribute most to the results of the study. For analysis of temporal and spatial aspects of epidemics or for comparisons across pathosystems, the components of the monocycle may yield important information to explain the behaviour of diseases in time and space. In the following sections, criteria of the components listed in Table 4.6 will be examined. This information can be used in the comparisons of Chapters 5–7. Most of these criteria are studied under controlled conditions (Sect. 3.2.1.2), but the results obtained can be validated in critical field experiments.

Such comparative field experiments, for instance, were conducted by Mouliom-Pefoura et al. (1996) with *Mycosphaerella fijiensis* and *M. musicola* (black and yellow Sigatoka diseases) of *Musa* spp. at three altitudes in the Cameroons. The components of the infection chain interacted and may have compensated for each other. At low altitudes (80 m above sea level) infection and incubation periods of both pathosystems did not differ. However, the expansion of lesions lasted longer for black than for yellow Sigatoka, which might explain the disappearance of the latter disease in the lowlands. However, at 900 and 1350 m, conidia of *M. musicola* developed faster than in the other pathosystem. In addition, the incubation period was somewhat shorter for *M. musicola* (17–20 days) than for *M. fijiensis* (22–24 days). All of these factors favour the prevalence of *M. musicola* at higher altitudes in this part of the tropics.

4.3.2.1 Infection

Infection is a process that develops in several physiological states and visible morphological structures (e.g. germ tubes, appressoria, haustoria). Criteria of interest in epidemiology are the duration of the infection period, the time spell during which a pathogen is present on a susceptible host and conditions are favourable for infection, the infection time (e.g. in hours) needed for a pathogen to penetrate its host, and the infection and disease efficiency (Table 4.6). During the infection period, pathogens are exposed to negative environmental factors, control measures, and to host defence reactions. A short infection time obviates these adverse factors more easily than a longer one. The effects of these factors first result in infection efficiency, and later in disease efficiency. Infection efficiency (Table 4.6) is defined as the infections that result from an inoculum of 100 propagules, under optimal temperature and non-limiting wetness duration. Successful infections can either be seen microscopically (e.g. appressoria formed) or visually (e.g. the streaks of yellow Sigatoka disease). The parameter for disease efficiency is the proportion (or percentage) of manifest lesions per 100 propagules as above. The difference between infection efficiency and disease efficiency is that not every infection develops into a lesion which then produces inoculum. In field experiments with the yellow Sigatoka disease of banana 18–69% of the yellow streaks (infection efficiency) turned into brown lesions (disease efficiency) depending on leaf insertions and weather (Kranz 1968d).

Many studies have been published on the infection process and the factors that affect it; a few examples are presented here. For *Puccinia recondita* de Vallavieille-Pope et al. (1995) found a disease efficiency 12 times greater than for *P. striiformis*. The range of temperatures suitable for infection periods was 5–25 °C for *P. recondita*, but only 3–12 °C for *P. striiformis*. The minimum continuous dew period required for infection increased from 4–6 h at optimal temperatures (8 °C for *P. striiformis* and 15 °C for *P. recondita*) to at least 16 h at suboptimal temperatures. The percentage of infection as the function of the duration of the continuous dew period was described by a Bertalanffy-Richards function with temperature-dependent parameters. A negative exponential function fitted the duration of wet periods necessary for infection. Infection efficiency was not affected by dry periods after the minimal continuous wetness and was not different to infection under continuous wetness. Between leaf and stripe rust on wheat (*Puccinia recondita* and *P. striiformis*), the infection efficiency was up to 40% for *P. recondita* and only 5% for *P. striiformis*, but this was compensated by sporulation resulting in a progeny/parent ratio three times higher for *P. striiformis* (Sache and de Vallavieille-Pope 1993).

Infections of four pathogens that cause plant diseases from semi-arid habitats were studied in growth chambers at various temperatures and durations of wetness periods (Bashi and Rotem 1974). *Stemphylium botryosum* f. sp. *lycopersici* infected tomatoes equally well after either a long wetting period or after several short moisture periods interrupted by dry spells. With a fast penetration (a short infection time) of potato leaves, *Phytophthora infestans* compensated for the sensitivity of sporangia to desiccation. Germ tubes from conidia of *Alternaria solani* also penetrated potato leaves rapidly, although they are drought resistant. However, infection efficiency of *A. solani* on potato was higher after a continuous wetness period rather then when they were exposed to a few short periods. *Uromyces appendiculatus* in beans developed in semi-arid habitats only in humid seasons because of the comparatively slow infection (long infection time) and low survivability of urediniospores between short wetness periods.

4.3.2.2 Incubation and Latent Periods

Incubation periods, as part of latent periods, last from infection or inoculation to lesion manifestation. With a certain delay in most diseases, inoculum formation in the lesions begins. This marks the end of the latent periods p and the beginning of the infectious period i. The duration of latent period p (in hours, days, weeks, etc.) determines the number of pathogen generations. Thus, the latent period affects the upper limit of explosiveness of an epidemic. The more generations a pathogen has in the growth period of hosts allows more inoculum to be produced by the pathogen and, therefore, the more intensively a disease can build up. Virologists (Thresh 1983) may lump together the terms incubation and latent period as the spell of time for symptoms to develop, because in virus diseases, the infectious period already starts during pathogenesis. Virus particles may be disseminated earlier by vectors (including man), that is before symptoms are seen, than what occurs in most bacterial and fungal diseases. Such extended infectious periods i, in fact, shorten the latent periods p (generation time), and, theoretically, could enhance epidemics. There is no evidence that virus diseases, in general, progress more rapidly or are more severe than diseases caused by fungi or bacteria.

Parameter values for incubation periods may be estimated during ongoing epidemics, albeit with some effort. This monitoring can be done by tagging leaves, recording their emergence, and spore catches on leaves nearby, and relate them to temperature, leaf wetness periods, rain events, etc. The ranges of incubation and latent periods for some pathogens obtained in various unrelated experiments are compared in Table 4.11.

Latent periods (the incubation periods inclusive) react most sensitively to temperature, the level of quantitative resistance, plant age (Fig. 4.7), density of infecting inoculum, nutritional status and other factors that affect the predisposition of host plants. The latent periods of *Erysiphe graminis* f.sp. *hordei* on spring barley and *Stagonospora nodorum* winter wheat were compared in growth chamber experiments (Fig. 4.7). The latent periods of both these diseases follow an opposite trend. Due to juvenile resistance in wheat to *S. nodorum*, latent periods on younger leaves are longer than later in the season when susceptibility increases. In barley it is the reverse as young leaves are susceptible while the older leaves progressively develop adult plant resistance to *E. graminis*, and thus the duration of the latent period is extended. Aust and Hau (1983) calculated that for *S. nodorum* 33% of the variability of the length in latent periods was due to changes in resistance, and 16% to temperature.

Table 4.11. Range of duration (in days) of incubation and latent periods of pathosystems on six to eight cultivars of their host plants. (Adapted from Rapilly 1991)

Stagonospora nodorum on wheat (Rapilly 1977)	
Incubation period	4.5–6.5
Latent period	6.8–8.2
Spilocea pomi on apples (Amorim 1984)	
Incubation period	5–14
Latent period	8–19
Puccinia hordei on barley (Johnson and Wilcoxson 1978)	
Latent period	7–13
Pyricularia grisea on rice (Yeh and Bonman 1986)	
Latent period	5.7–6.4
Mycosphaerella fijiensis on banana (Fouré 1982)	
Incubation period	13.7–25.4
Latent period	17.0–45.0
Mycosphaerella musicola on banana (Kranz 1968d)	
Latent period	28–79
	28–45[a]

[a]During months favourable to infection.

How pre-inoculated Bean Line Pattern Mosaic Virus (BLPMV) and temperature during incubation or latent period affected the pathosystems bean – *Uromyces appendiculatus* (rust), and bean – *Phaeoisariopsis griseola* (angular leaf spot) on two bean cultivars was studied in growth chambers (Bassanezi et al. 1998). The following parameters were measured: the development rate of lesion density, lesion size and disease severity. Temperature effects on monocyclic components of these pathosystems followed optimum curves that were fit by the generalised beta function (Fig. 4.8; Hau 1988; Hau and Kranz 1990). Rates for the development of rust were higher than for angular leaf spot. There was no effect of the virus pre-infection on the latent periods of either fungal disease on either cultivar.

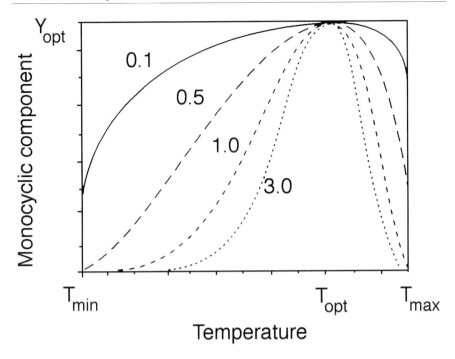

Fig. 4.8. Graphic representation for four values chosen for the shape parameter B_3 of the generalised beta function. T_{min} and T_{max} are the lowest and highest temperature for the development of a given monocyclic component. T_{opt} is the optimal temperature, and Y_{opt} the corresponding maximum disease response. (Bassanezi et al. 1998)

4.3.2.3 Lesions

After the incubation period, lesions become manifested and can be measured in numbers/unit host, size and age. Lesions determine the disease intensity (see Kranz 1988a; Hau et al. 1989; Nutter et al. 1993b; Francl and Neher 1997). Intensity covers both disease incidence and severity. Disease incidence (%) is the proportion of diseased sampling units, e.g. leaves, plants, fields in a sample (usually irrespective of their individual severity, but a certain threshold of severity, e.g. ≥3 lesions per leaf, may be implied). Disease severity (%) is the percentage of the susceptible host tissue in the sampling unit covered by lesions that results from both the number of lesions/unit host area and the lesion size (e.g. mm^2). A common practical criterion for severity is the estimation of percent host surface covered by lesions in a sample. Many small lesions are an indication that either inoculum or disease efficiency was high while a few large lesions are rather from low inoculum and the plant's high susceptibility. There are also differences between pathosystems. An increase in the number of lesions was higher in leaf rust than in stripe rust of wheat because of a latent period 2 days shorter than that of leaf rust (Sache and de Vallavieille-Pope 1993).

Lesion expansion can enhance severity without any additional new lesions and thus, is an important epidemic component (Hau 1988). Berger et al. (1997), after an extensive, quantitative review of lesion expansion in several pathosystems, stress that "...lesion expansion is such an important portion of the total disease syndrome of many polycyclic epidemics, the reduction of lesion expansion should be emphasised to manage these pathosystems." From the epidemiological point of view, the lesions proper, i.e. the potentially sporulating symptoms of the disease, have to be distinguished from decaying plant parts as a consequence of disease. For assessment of yield loss and decision making, lesions and decay may be different and have to be assessed accordingly (see Sect. 7.1).

Vitti et al. (1995a) in a comparative study under a controlled environment on the monocycles of *Puccinia sorghi* and *Exserohilum turcicum* on maize, showed that *E. turcicum* had a 15-fold faster lesion expansion rate than maize rust, which has a higher disease efficiency, at the end of the 10th monocycle, the y_{max} of rust was 12% higher compared to 5% for the leaf spot. However, after four polycyclic epidemics of both diseases that lasted 34–70 days (Vitti et al. 1995b), the number of lesions per cm^2 was 7- to 100-fold greater for rust, while the final disease severity was 3- to 18-fold greater for *E. turcicum*. Bassanezi et al. (1998, see above) found temperature optima for disease severity of angular leaf spot of beans ranged from 24.2 to 28.3 °C, and from 15.9 to 18.5 °C for bean rust. This relationship to temperature was not affected by the Bean Line Pattern Mosaic virus (BLPMV). The lesion density of rust and anthracnose on two bean cultivars 17 days after inoculation with BLPMV was significantly reduced 41 and 52% for rust and 17 and 29% for angular leaf spot.

Lesions may change their aspects during development and age in the course of an epidemic and with leaf age. This can affect spore production. An important aspect is the criterion "age groups of lesions and their proportions in host populations" which affects the power behind the dynamics of epidemics. Young lesions have a higher sporulation intensity than older lesions which eventually drop out of the reproduction, e.g. when necrotic or removed (see Table 4.12).

Table 4.12. Classification of powdery mildew lesions by age characterised by the colour of colonies, and the mean proportional sporulation intensity per lesion of *Erysiphe graminis* f. sp. *hordei*. (Rangkuty 1984)

Colour[a]	Conidia/lesion/day	Proportional sporulation
White	318.6±92.7	0.09
Light brown	3373.5±887.7	1[b]
Yellow-brown	1833.1±394.4	0.54
Dark brown	215.0±57.1	0.06
Blackish-brown	59.8±26.6	0.01

[a]Colour of colonies of barley powdery mildew. The youngest colonies are white and turn darker with age. The change must not be correlated with time in days as host plant resistance, position of plant organ and weather factors can have an effect.
[b]The maximum mean sporulation which regularly occurred with this shade of colour is set equal to 1, and all the other classes of colour relative to it.

4.3.2.4 The Infectious Period

The infectious period i begins when the latent period ends. The duration of i also exerts a great influence on epidemics (e.g. Van der Plank 1963; Zadoks 1971; Luo and Zeng 1995). A measurement term for i is its duration (in days or any other time scale) for an individual lesion. During the infectious period, inoculum is the amount of infectious propagules in lesions where it is formed, in transit when being dispersed, and after landing on healthy plant parts for new infections. Parameters for inoculum in lesions are sporulation intensity and capacity. Sporulation intensity is measured as propagules/mm^2 lesion/day, or as spore catches in spore traps (number of spores/time unit). The sporulation capacity of a lesion is the integral of its changing sporulation intensity (Table 4.12) over the infectious period i, expressed in numbers M per lesion. Sporulation potential here is understood as the assessed (or assumed) capability of a population of lesions in an area (plot, field, etc.) to produce inoculum at time t_i as a function of the mean age of lesions in a pathogen population. Its parameter is (mean age) $\times$ (M_t) which could describe the potential that an epidemic has at t_i. This parameter is important for disease management, but it is difficult to assess since it depends on the mean number of lesions present at t_i and the mean age of the population of lesions. A solution to this problem may come from simulators based on estimates for the relevant variables, that is, the proportion of age groups of lesions and their sporulation intensities. Spore density or inoculum at time t_i of an epidemic is the amount of propagules measured by means of spore trapping devices in a given area (plot, field etc.). The effectivity N×E (Zadoks and Schein 1979) of this inoculum is affected by the mean age of the propagules upon arrival on a host plant, its susceptibility, weather etc.

Young lesions caused by many fungal pathogens tend to sporulate more profusely than older ones. With increasing age, sporulation intensity fades and finally stops. A high proportion of young lesions thus produces more inoculum which can result in a more rapid development of epidemics, i.e. faster rates r. Lesion age in some pathosystems may be distinguished morphologically, e.g. by their colour. If the various stages are distinct, the rhythm of sporulation can be observed. In such a case, the number of spore crops during i can be another parameter for comparison. Rangkuty (1984) defined five age classes for barley powdery mildew by lesion colour (Table 4.12). Pinnschmidt et al. (1995b) used nine classes of rice leaf blast lesions and five classes for the panicle blast by a combination of changing colour in the lesions and their border lines.

Inoculum in situ on host surfaces can be counted with the aid of a microscope on films of dried transparent glues (or simply nail varnish) stripped from the epidermis. However, a coefficient to calculate true densities of spores may have to be established, as this method usually fails to collect all spores. For conidia of powdery mildew on wheat and barley, the spores counted were multiplied by 1.333 to adjust the counts to the actual number of spores present (Merchán and Kranz 1986).

Like the latent period p, the infectious period i is closely related to cultivar resistance. The more resistant the cultivar, the shorter is i. A major factor that affects the duration of the infectious period and sporulation intensity is temperature. This temperature effect applies for a given level of resistance, as long as humidity or leaf wetness is sufficient. For example, *Rhynchosporium secalis* and *Stagonospora nodorum* produced more conidia, when the wetness period was extended from 8 to 16 h per night and the night/day temperature was elevated from 10/20 to 15/25 or 20/30 °C (Bashi and Rotem 1975). Both liquid water and dryness may reduce the inoculum of bacteria and fungi on plants. All of these parameters as criteria are variables among pathogens, or aspects within pathogens, like infection, latent periods etc. The effects of host and environment on sporulation in vivo are well reviewed by Rotem et al. (1978). The differences of sporulation intensity were rather distinct among the three pathogens that cause downy mildew of maize, viz. *Peronosclerospora sorghi*, *P. sacchari* and *P. philippensis* that originated from various countries (Duck et al. 1987). Light affects inoculum formation differently, Hau and de Vallavieille-Pope (1998) cite Clifford and Harris (1981) who found that low light intensity curtailed the rate of lesion development and spore production of *P. recondita* f. sp. *tritici*. Contrarily, disease severity of barley powdery mildew (*Erysiphe graminis*), was significantly favoured by a few hours of shade per day (Kranz and Aust 1979).

4.3.2.5 Dispersal of Inoculum

Pathogens are dispersed by contact, wind, water, vectors, and by man through seed and infected plant material. This occurs in three distinct phases: liberation, transit (flight) and landing. Spore liberation at the start of dispersal can be compared either by descriptive terms (e.g. through shock, wind eddies, water and by vectors) or quantitative terms when parameter values for the liberating force are measured. These various vehicles of dispersal have implications on the distance and efficacy of disease spread. In addition, the effect of barriers to spread differs among wind-, vector- and splashborne propagules. A large proportion of fungal pathogens has adapted to physical factors of airborne and splash dispersal. Bacteria are rather waterborne unless they are dispersed in windborne droplets. Most viral pathogens are transmitted by vectors (Table 4.5). Comparative studies on liberation and dispersal of various species of fungal pathogens have been reviewed, for instance, by Fitt and McCartney (1986) and McCartney and Fitt (1998). A comparison of rain effects on splash dispersal was done with three species of *Colletotrichum* in the laboratory (Ntahimpera et al. 1999). This experiment elucidated the effect of rain impaction on liberation and the distance of primary dispersal. The conidia of the three *Colletotrichum* spp. that infect strawberry were similarly, but not identically, dispersed by rain splash and differed in some of the parameters measured. Sections 6.1–6.3 have more about dispersal.

Inoculum in transit by air follows diurnal curves and can be measured quantitatively by means of appropriate traps, e.g. number spores related to volume air (m^3), trapping area (cm^2), and time (h). Gregory (1961) compared diurnal patterns and distinguished groups of diurnal periodicity (Fig. 4.9).

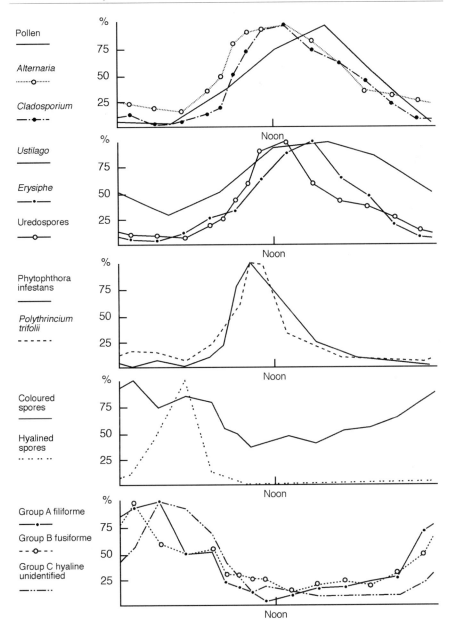

Fig. 4.9. Five groups of diurnal periodicity patterns of inoculum dispersal expressed as a percentage of the peak geometric concentration recorded in summer 1952 with the Hirst spore trap and partly corrected with data from tests in wind-tunnel using *Lycopodium* and *Ustilago* spores (Gregory 1961). *Topmost two groups* afternoon pattern, *third group* forenoon pattern, *fourth group* basidiospores with mainly nocturnal dispersal and *fifth group* nocturnal pattern (notably balistospores of *Sporobolomyces*, *Tilletiopsis*, and basidiospores of hymenomycetes)

The flight distance (in metres, kilometres) covered by dispersed inoculum can be compared between types and size of propagules and vehicle of transport, as well as flight or drift of vectors. For windborne dispersal, the following factors affect the velocity of sedimentation c_e, for a given spore size: wind speed u, air mass A, and local turbulences. These factors also affect the distance covered during the flight (Table 4.13). No differences were found between the splashborne needle-like conidia of *Pseudocercosporella herpotrichoides* and the cylindrical spores of *Pyrenopeziza brassicae* dispersed in controlled experiments (Fatemi and Fitt 1983). The prevalent form of dispersal will distinctly determine focal and long distance spread of diseases (see Sects. 6.3 and 6.4). Special conditions prevail for the dispersal of soilborne diseases (Table 4.4). Factors that affect short and long distance transport and the gradients are being dealt with in Section 6.

The last stage of dispersal is the landing of inoculum. Vectorborne transmission of inoculum has a higher probability of hitting a target host than spores carried by the wind. The density of inoculum is measured on the host plants (see p. 75), or with trap plants (e. g. seedlings) inside the crop. Spore traps which catch inoculum alive, like the jet-spore trap (Schwarzbach 1979), allow for a check of the viability of the landed inoculum. An infective density of inoculum may follow an optimum curve in its efficiency. If the inoculum density on host tissue is too high, it may decrease the disease efficiency, e.g. *Erysiphe graminis* f. sp. *hordei* and *Puccinia striiformis* f. sp. *tritici*.

Table 4.13. Probable[a] flight range (km) of spores depending on their air mass exchange coefficient (A), velocity of spore sedimentation (c_e, here for ellipsoid spores) and wind speed (u). (Schrödter 1987)

A (g/cm/s)	c_e (cm/s)	u (m/s) 2	4	6	8	10
10	0.138	800	1600	2400	3200	4000
	0.975	16	32	48	64	80
	1.300[b]	9	18	27	36	45
20	0.138	1600	3200	4800	6400	8000
	0.975	32	64	96	128	160
	1.300	18	36	54	72	90
50	0.138	4000	8000	12,000	16,000	20,000
	0.975	80	160	240	320	400
	1.300	45	90	135	180	223

[a]The probable flight range is where 50% of the inoculum in transit has been deposited.
[b]The value of 1.3 for c_e applies to the sporangia of *Phytophthora infestans*.

Latent periods were shorter and lesion size was smaller with high inoculum density of *E. graminis* f. sp. *hordei* and *P. recondita* f. sp. *tritici*. Sporulation intensity decreased with the increase in density of lesions in powdery mildew and wheat stem rust (*P. graminis* f. sp. *tritici*), whereas sporulation of *P. recondita* f. sp. *tritici* was relatively independent of lesion density (Hau and de Vallavieille-Pope 1998).

4.3.2.6 Survival

An unfavourable environment (e.g. low or excess temperature, dryness) or the absence of host plants force pathogens to develop strategies to ensure survival for shorter or longer periods. Although it is part of the life cycle, survival, however, may occur in some components of the infection chain, e.g. in lesions and as inoculum already formed. Two quantitative criteria, viz. surviving inoculum Q and longevity of inoculum, and a descriptive one, survival strategy, are suitable for comparison across pathogens and epidemics. Dickinson and Lucas (1977; Fig. 4.10) distinguish between various means of inactive and active survival. For the surviving inoculum Q before primary infection, a number of quantitative assessment methods are known for a particular pathogen. Benson (1994) gives an account for soilborne pathogens. For pathogens that hibernate on above-ground plant parts, an appropriate sampling at relevant times depends on the site, e.g. fruits, bark, debris. Subsequent analyses in the laboratory use established methods and criteria. The parameter of longevity is in the appropriate time unit (weeks, years) for various pathogens and types of propagules (Fig. 4.11).

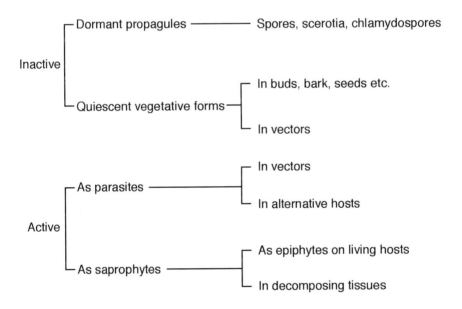

Fig. 4.10. Strategies adopted for survival by plant pathogens. (Adapted from Dickinson and Lucas 1977)

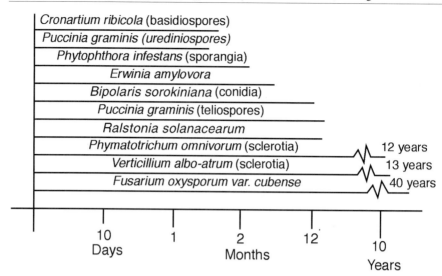

Fig. 4.11. Longevity (survival periods) of some pathogens in the field. (Adapted from Dickinson and Lucas 1977)

4.4 Epidemic Competence of Pathogens

The equivalence theorem of Van der Plank (1963) states that more favourable weather conditions, less host resistance, and greater pathogenicity are equivalent in their epidemiological effects. For the role of pathogens, however, not only pathogenicity determines the epidemic competence of a pathogen to cause more or less severe epidemics. Additional characteristics of the pathogen to contribute to their epidemic competence are given in Table 4.14. Epidemic competence of pathosystems thus can be due to different capabilities. For example, the epidemic competence of powdery mildew, stem and leaf rust of wheat rests on a fast infection cycle, large sporulation capacity and a high disease efficiency. The slower infection cycle of stripe rust is compensated by lesion growth, which also affects powdery mildew (Hau and de Vallavieille-Pope 1998). Tropical rusts, therefore, occur in a relatively stable environment with a nearly permanent availability of susceptible host tissue. Tropical rusts have a longer infectious period i, a later sporulation peak and a lower disease efficiency than some temperate rusts (Sache and de Vallavieille-Pope 1995).

Table 4.14. Characteristics of epidemic competence of a pathogen within a given patho-system

Characteristic	Epidemic competence is rather high when
Pathogenicity	High[a]
Relative parasitic fitness	High
Dispersal of inoculum	Easy and over larger distances[b]
Survivability	No limiting factor for primary infection
Environmental requirements	Wide favourable ranges in factors affecting

[a]See Section 4.4.1 and Tables 4.15 and 4.16.
[b]Inoculum can be liberated easily and carried over large distances.

4.4.1 Criteria for Pathogenicity

The term pathogenicity comprises (1) virulence, (2) aggressiveness, and (3) fitness of pathogens (sensu Van der Plank 1963). Virulence in the gene-for-gene-hypothesis is the qualitative capability (yes-or-no reaction) of a pathogen genotype to recognise and infect a certain resistance genotype of the host plant. The biological units by means of which virulence genes of pathogens interact with host genotypes are the fungal spore, bacterial cells or viral genomes. A pathotype carries one or several virulence genes. These are identified by matching resistance genes in so-called differentials (see Sect. 7.3.2), or by molecular analysis of genes. However, each fungal spore in addition to its virulence genes (if they exist) has other genes which are needed for the survival of the genotype in a changeable environment. These additional genes enable the pathogen to adapt to ranges of temperature, and fungicide resistance, for instance (Sect. 7.4.2).

Pathogenicity in populations of the pathogen is subjected to selection processes on genotypes as subpopulations, or metapopulations, e. g. pathotypes or races (Park et al. 2001) through both qualitative and quantitative resistance (p. 55). This ensures survival of the pathogen species in the course of micro-evolution. The pathogenicity requires virulence as a qualitative characteristic and measurement terms as criteria for aggressiveness as quantitative parameters (Table 4.15). Some of these parameters appear again in Table 4.16 as criteria for parasitic fitness.

Table 4.15. Variables (parameters) to measure whether aggressiveness of the pathogen is lower or higher when tested with standardised inoculum

Variables	Aggressiveness	
	Low	High
Disease intensity (either y_o, y_t or y_{max})[a]	Low	High
Incubation period	Long	Short
Latent period p	Long	Short
Infection efficiency	Low	High
Disease efficiency	Low	High
Infection rate r	Low	High
Lesion size	Small	Large
Infectious period i	Short	Long
Sporulation intensity	Low	High

[a]Initial disease intensity y_o, maximum disease intensity y_{max}, and y_t, the disease intensity measured at any time t during the epidemic. Disease intensity may be either incidence or severity.

Table 4.16. Absolute fitness variables

Variable	Fitness is higher when
Infection time	Short
Infection efficiency[a]	High
Duration of latent period p	Short
Disease efficiency[b]	High
Disease intensity[c]	High
Duration of infectious period i	Long
First spore crop	Large
Number of spore crops during i[d]	High
Sporulation intensity/capacity[e]	High
Viability of propagules	Long
Environmental requirements	Less stringent
Compensatory capabilities	More flexible
Multiplication rates	High

[a] Number of visible infections not yet developed into lesions (for example, the yellow streaks of the Sigatoka disease of banana) per cm^2 from 100 propagules inoculated per unit area host.

[b] Number of lesions per 100 propagules inoculated per unit area host. See also Royer and Nelson (1981).

[c] Disease intensity (either incidence or severity) as parameter for aggressiveness when the amount of infective inoculum on the host is not known.

[d] Spore crop, the quantity of spores etc. formed when resporulating after previous spores were liberated and dispersed.

[e] Sporulation intensity = number of propagules/unit lesion area/unit time; sporulation capacity = total inoculum produced per lesion during the infectious period i. The classical concept of fitness in this context is more offspring.

The matching virulence genes are selected in qualitative resistance. In quantitative resistance, pathogen genotypes with a higher aggressiveness are selected to become dominant in a population. Selection processes may be slow or rather rapid depending on pathogenicity (Table 4.15), parasitic fitness (Table 4.16), and the number of generations (latent periods) produced during a vegetation period of a host plant. Epidemics become more serious with an increase in more pathogenic genotypes which then replace the less pathogenic ones. The major driving force in selection of the pathotype that affects the strength of the ongoing process is the relative parasitic fitness supported by the popularity of resistant host genotypes (p. 53).

Measurement terms of absolute parasitic fitness parameters in Table 4.16 are usually obtained in laboratory or growth chamber experiments. Relative fitness is also measured in growth chambers, usually over a number of pathogen generations as a time scale in comparison with a standard pathotype (Fig. 4.12).

The relative fitness of races or isolates (when assumed constant) can be compared to a standard when computed in the equation below with the following parameters (see Welz 1988): W_q is the fitness of race q, and $W_p=1.0$ for the reference race, with their proportions in equation:

$$q_n/p_n=(q_0/p_0)W_q{}^n.$$

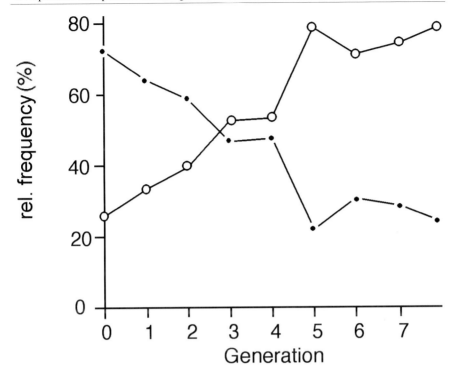

Fig. 4.12. The change in the proportions of two races of *Erysiphe graminis* f. sp. *hordei* due to their relative fitness in the course of eight generations starting at a ratio 25:75% in the mixture of conidia. (Stähle et al. 1984)

If the proportions of q_0 and p_0 are recorded at the start, and q_n and p_n after n generations (at least more than twice, better five times or more), a linear regression can be computed as y=a+bx with a slope of $\ln W_q$. Thus $W_q = e^b$.

$$\ln(q_n/p_n) = \ln(q_0/p_0) + n \times \ln(W_q)$$

However, Østergard and Shaw (1996) doubt that linear regressions are appropriate to estimate fitness by this type of assessment because of allo-infection and an intrinsically non-linear nature (Østergard and Hovmøller 1991). This procedure will require explicit likelihood- or resampling-based methods.

Absolute and relative parasitic fitness for a given race-cultivar combination can best be compared under controlled conditions. Measurement and comparison of fitness for an entire pathogen population of many races of different fitness under field conditions are very difficult. Normally, fitness is related to generation time, the duration of which is difficult to be assessed in the course of an epidemic under field conditions. In order to cope with this problem, Hau (1990) estimated a mean generation time from on-site recorded temperature by means of the equation

$$p = K(T-2-N(30-T)-M \text{ if } 2 < T < 30 \text{ °C}$$

where the latent period p of *Erysiphe graminis* f. sp. *hordei* is a generalised beta function of temperature (T); K, N and M are constants (shape parameters) known from studies under a controlled environment. The three equations above do not account for the effects of the density of lesions and the number of propagules produced by them. This shortcoming for the comparison of density effects on the selection of individual pathotypes may be overcome by a selection coefficient $C(f_i)$ (Kranz 1987), which is based on the apparent infection rate r (Van der Plank 1963):

$$C(f_i)=1/(t_2-t_1)[\text{logit}(y_2 f_i)-\text{logit }(y_1 f_i)]$$

This equation requires both the disease intensities y_1 and y_2 at two subsequent dates t_1 and t_2 as inputs. The disease intensities provide an estimate of the change in densities of lesions (and, indirectly, the number of spores), and frequencies of races f_i measured at these dates. This equation estimates the effect of changing disease intensity on the frequency for one of the races sampled at these dates. $C(f_i)$ may be calculated for other races f_i identified in the sample using the same values of y. From the changes in race frequencies within the sample between the two dates of disease assessment, conclusions may be drawn about the effect of pathogen density on the relative parasitic fitness of races. The function $C(f_i)$ does not need any estimates or assumptions of generation time.

Pathotype composition of a pathogen population as a consequence of selection can be compared by criteria in Table 4.17 for the measurements of their variables (see pp. 80–82).

Table 4.17. Terms in use[a] as criteria for comparisons of the states and compositions of pathogen populations

Pathogenic diversity	Richness of a pathogen population in virulence genes or races (genotypes), and evenness of their distribution
Evenness	Degree of even distribution of genotypes in a pathogen population[b]
Richness	Number of genotypes in pathogen population[b]
Genetic similarity	Number of characters or alleles shared by two isolates relative to the total number of characters or alleles of the two isolates. Measured e.g. by the Dice index[c]
Frequency	Relative proportion (%) of a genotype in the pathogen population
Flexibility	The adaptability of a race or a pathogen population as a whole due to its genetic diversity
Virulence association	The virulence genes combined in a race or in a regional pathogen population to a greater extent than expected, due to linkage or hitchhiking selection
Gene flow	The movement of genes or genotypes among pathogen populations

[a]Compiled from Wolfe and Schwarzbach (1978), Welz (1988), McDermott and McDonald (1993).
[b]This criterion, and diversity, can be measured by the Shannon index, or a similar index.
[c]Dice index (see Sneath and Sokal 1973).

Among the criteria in Table 4.17, richness, genetic similarity and frequency are rather straightforward terms. Flexibility and gene flow are somewhat descriptive. The criterion "diversity" measures genetic variations of a population or sample. The most common indices are the Simpson's measure of diversity $D=1-\Sigma n_i(n_i-1)/n(n-1)$ with n_i the collected number of the i-th phenotype, n the sample size, and the Shannon index $H'=-\Sigma p_i \ln p_i$ with p_i as the frequency of the i-th phenotype. The Shannon index essentially reflects the richness of a population and the evenness of the frequency distribution. Critical aspects for the application of the two indices are discussed by Welz (1988). For a comparison of virulence associations in genotypes of a pathogen population, we refer to Welz (1988) who also recommends the G-test of independence as being more appropriate than the Chi^2 method.

Diversity of pathotypes tends to be greater in sexually reproducing than in asexual populations with dominant pathotypes. Sexual progeny as primary inoculum increases the frequency of pathotypes at the start of a new growing season. Selection during the season then acts against the lesser fit pathotypes under the prevailing conditions. This selection process was demonstrated for wheat stem rust by Groth and Roelfs (1982) and barley powdery mildew by Welz and Kranz (1987). Pathogens with frequent sexual recombination may have less complex races, but a higher diversity, i.e. more pathotypes (Hau and de Vallavieille-Pope 1998). In contrast, the complexity of races (i.e. more virulence genes) increases when usually more prolific asexual spores face the selection pressure of race-specific host resistance genes for a longer period. Facultative parasitic pathogens may develop a saprophytic fitness during a saprophytic life. This may affect the race frequency in the aerospora and alter the pathotype pattern of a population at the time of primary infection. A comparison between related species revealed that parasitic and saprophytic fitness were negatively correlated (Leonard et al. 1988): *Cochliobolus heterostrophus* (anamorph *Bipolaris maydis*), which causes southern corn leaf blight (SCLB), has high parasitic fitness as indicated by its rate r. This fitness occurs rather late in the season and starts at low levels of y_0. *C. carbonum* (anamorph *B. zeicola*), the incitant of carbonum corn leaf spot (CCLS), occurs early with high values of y_0, reaches only low r-values during the epidemic and the disease attains low y_{max} only.

Weather factors also affect the selection of pathotypes. In a field experiment over 3 years, Ohl (1991) compared, relative to the then prevalent race 23, positive and negative reactions of pathotypes of *Erysiphe graminis* f. sp. *hordei* to temperature, humidity, leaf wetness duration and rainfall. Even small differences in fitness among the ten pathotypes may lead to differential reactions to prevailing weather (Table 4.18), which in the long run could change race frequencies(see Sect. 4.4.2). For instance, pathotype 15 was prevalent under the more humid conditions of northern Germany. In Ohl's experiments a positive correlation was shown in some pathotypes with high frequencies of leaf wetness duration and rainfall when temperature was neither too high nor too low.

Table 4.18. A frequency correlation analysis of the relative fitness of pathotypes of *Erysiphe graminis* on spring barley that reacted either negatively (−), positively (+)[a], or intermediately (±) to weather factors as compared to pathotype 23 as the standard. (Ohl 1991)

Name of Pathotype[b]	Weather factor							
	Temperature		Relative humidity		Wetness (h)		Rainfall	
	Low	High	Low	High	Low	High	Low	High
7	−	+	nc[c]		±[d]		−	+
15	−	+	±		±		±	
31	+	−	−	+	−	+	±	
39	−	+	±		nc		+	−
55	−	+	−	+	±		nc	
71	+	−	−	+	−	+	−	+
87	+	−	±		−	+	±	
135	−	+	−	+	−	+	−	+
151	−	+	nc		+	−	+	−
183	−	+	+	−	+	−	+	−

[a]On cv. Villa (with the ineffective resistance gene Mg and corresponding virulence gene Vg) relative to pathotype 23 (with virulence genes Vg, Va6, Va12, Vh) as standard.
[b]The figures as names of the pathotypes are from the British Mildew Differential (see Table 4.19).
[c]No clear reaction.
[d]Pathotypes with an intermediate reaction at the optimum of the relationship for the mean of the weather factor

Cold temperatures during winter and the passage through winter barley which practically had no effective resistance genes for the virulences of spring barley (Table 4.19) caused a reduction of disease severity and altered pathotype frequencies in the population.

Table 4.19. Pathotype frequencies (%) of *Erysiphe graminis* f. sp. *hordei* before and after winter 1988/1989.[a] (Welz and Kranz 1997)

Pathotype Name[b]	Virulence genes[c]	Dates of sampling	
		30 Dec 1988	2 May 1989
15	1.2.3	9.1	22.1
23	1.2.3.−.5	10.7	24.7
31	1.2.3.4.5	2.5	7.4
47	1.2.3.4.−.6	15.7	2.5
55	1.2.3.−.5.6	9.1	2.5
63	1.2.3.4.5.6	9.9	2.5
103	1.2.3.−.−.6.7	5.8	7.4
Rest[d]		37.2	31.0
No. of pathotypes	25	33	
Shannon index	2.77	2.67	

[a]Results in a second year were practically the same.
[b]Using the nomenclature of Habgood (1970).

[c]Numbers refer to sequence of differentials (R-genes) in the British Mildew Differentials to which pathotypes have matching virulence.
[d]The "rest" comprises 18 identified pathotypes.

The frequencies of races 15, 23 and 31 in Table 4.19 indicated a high degree of adaptation to prevailing conditions. In addition, most virulence gene frequencies present in the pathotypes were also significantly different in a test of homogeneity (Chi^2=67.87, df=5, P<0.001), as were the two pathotype frequency distributions before and after the winter, though the Shannon index of diversity remained unchanged (t=0.680, df=213, P<0.001).

The selection in pathogen populations for fungicide resistance practically follows the same processes as for virulence (see p. 82). Only two additional parameters are needed as criteria for comparison apart from fungicide sensitivity and insensitivity (not resistant, resistant). These are:

- ED_{50} or ED_{95}, the effect dose, killing 50 or 95% of the test organisms. This is the parameter for the degree of fungicide resistance.
- t_s, the standard selection time (Skylakakis 1982) as $t_s=1/(r_2-r_1)$, with r as the apparent infection rates for the resistant strain (r_2) and fungicide sensitive strain (r_1).

4.4.2 Compensation Within Epidemics

Disproportionately small quantitative deviations of parameter values that affect components of the monocycle (infection chain) can trigger large qualitative and quantitative changes in disease intensity and epidemic behaviour. This is due to compensating interactions among factors, states and phases, particularly when one or more conditions are suboptimal for an epidemic development (Rotem 1978). This phenomenon explains, for instance, why in an oasis of the Sahara Desert substantial rust severity was found on a wild herbaceous plant species (Kranz, unpubl. observ.).

Aust et al. (1980) elaborated on this concept of compensation and discussed four categories (Fig. 4.13):

1. Factor-for-factor: a highly favourable factor in a phase of the life cycle or in the infection chain compensates for a simultaneously unfavourable factor, e.g. a temperature that is marginal for infection is compensated by long wetting periods.
2. Strength-for-weakness: a specific weakness in a pathogen can be compensated by a specific strength, e.g. low inoculum production by high disease efficiency of propagules.
3. Phase-for-phase: a high frequency of a favouring factor in one phase of the life cycle can compensate for a low frequency of another phase, e.g. high survival of inoculum can ensure higher primary infection even under more unfavourable conditions for infection; abundant sporulation during favourable weather and its dispersal to compensate for infection in subsequent less favourable periods; frequent growth flushes of susceptible host tissue may compensate for erratic weather conditions for dispersal of inoculum.

4. Resistance-for-weather: ontogenetic changes to a higher susceptibility of the host can compensate for periods of unfavourable weather conditions for sporulation and infection.

In an experiment in growth chambers, the compensatory effect of temperature, inoculum density and leaf wetness duration (LWD) on the latent period of *Stagonospora (Septoria) nodorum* was studied by Aust and Hau (1981). Suboptimal temperature was compensated for by extended LWD or higher inoculum, and vice versa. In a multivariate regression analysis 45% of the variability in the duration of latent periods of *S. nodorum* could be explained by temperature, 12% by inoculum density and only 3% by the LWD, with 40% of the variability remaining unaccounted for (see p. 71).

Relatively low spore production by wheat leaf rust (*Puccinia recondita* f. sp. *tritici*) was compensated for by high disease efficiency of its urediniospores, whereas the very low disease efficiency of stripe rust (*P. striiformis*) was compensated for by high spore production, i.e. sporulation intensity. The number of lesions and the amount of spores produced per area of lesion on maize was higher for *Puccinia sorghi* than for *Exserohilum turcicum*. However, larger lesions of *E. turcicum* compensated for fewer lesions, and for this reason the y_{max} is higher (Vitti et al. 1995b). A favourable level of one factor, e.g. a high penetration rate of *P. recondita* compensated for a low survivability of germinated spores after a dry period during the infection period (de Vallavieille-Pope et al. 1995).

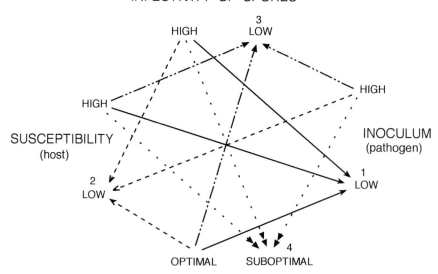

Fig. 4.13. Interactions between host, pathogen and environmental factors illustrating the general concept of compensation, e.g. low inoculum is compensated by high infectivity of inoculum or high susceptibility of hosts, low infectivity or low inoculum by optimal conditions for infection, etc. The disease triangle here becomes an octagon. (Aust et al. 1980)

Compensatory interaction of factors can also be observed in the field. An example to compare the effect of leaf wetness by sprinkler irrigation versus natural dew formation on two diseases of tomato (Table 4.20) under different environmental conditions is provided by Rotem and Palti (1969).

In Table 4.20, the effects of leaf wetness under given macroclimatic conditions differ between the two diseases. Supplementary moisture, especially dew, is required under marginal conditions of humidity for disease to develop. Sprinkling in the morning should be avoided as it favours pathogens with spores which do not survive a dry day and sunshine, like *P. infestans*. Instead, sprinkling before nightfall is recommended. Time of sprinkling does not matter to *A. solani* because of its drought-resistant conidia (Rotem and Palti 1969).

Table 4.20. Interaction of dew and sprinkler irrigation in their effects on *Alternaria solani* and *Phytophthora infestans* on potatoes and tomatoes under the environmental conditions in various localities in Israel. (Rotem and Palti 1969)

Macroclimatic conditions	Development of	
	A. solani	*P. infestans*
Completely arid and dewless oasis in the Negev Desert	Slight disease only under irrigation	No disease
Relative humidity during days <35% or less, nights rich in dew, no rain; N and W Negev	Dew alone suffices for epidemic development; no effect of sprinkling	Dew alone inadequate; blights require sprinkler
Relative humidity during days >60%, nights rich in dew, no rain; coastal plain	Dew alone suffices for epidemic development; no effect of sprinkling	Dew may suffice; sprinkling leads to outbreaks
Relative humidity always high, dew plentiful, no rain; coastal plain (spring/fall)	Dew alone suffices for epidemics; sprinkling without effect	Dew alone can support blight; sprinkling irrelevant

4.4.3 Interaction Among Epidemic Competent Pathogens in a Crop

Pathogens do not interact with host and external factors only, but also with other pathogens on the same host, e.g. two foliar diseases of wheat with and without the interference of root diseases (Table 4.21).

Results of interaction among pathogen as in Table 4.21 were further compared with a path correlation analysis (p. 90) for their effects on one yield component (Fig. 4.14). Apart from information on the interaction of diseases in a pest complex the weight of path coefficients D allow to design loss profiles and the identification of key pests for IPM schemes (see Sect. 7.1). These coefficients may be used as weights for variables and components as in Table 4.7.

Table 4.21. Mean disease severities (%) of *Erysiphe graminis* f. sp. *tritici* and *Stagono-spora nodorum* on wheat at growth stage DC 75, with and without incidence of *Gaeuman-nomyces graminis* (GG) and *Pseudocercosporella herpotrichoides* (PH) (summarised for one of the years from Jörg 1987)

	Mean disease severity (%) of							
	Erysiphe graminis				*Stagonospora nodorum*			
	GG		PH		GG		PH	
Organ	With	No	With	No	With	No	With	No
Ear	1	3	4	1	2	1	11	8
Flag leaf	4	11	2	1	47	25	17	25
Sec. leaf	13	27	4	3	22	11	45	47

In order to model interactions among pathogens, Weber (1996) employed ana-lytical models of modified Lotka-Volterra equations on epidemics of two patho-gens that competed on winter wheat, (i.e. the biotrophic Erysi*phe graminis* and the perthotrophic *Stagonospora nodorum*), studied concurrently for 2 years in field experiments. Two minimal interaction models first related total disease severity with their epidemic rates and capacity (maximum severity), together with interac-tion parameters. In addition, an extended interaction model described infectious and post-infectious compartments of *Stagonospora*. In all three interaction models *S. nodorum* competed with *E. graminis* for infection sites and, therefore, was linked to its density regulation term. The positive effect of *E. graminis* on *S. no-dorum* was modelled using two approaches: first, it is assumed that *E. graminis* increases the final severity of *S. nodorum*, and the rate parameter of the latter. In the second approach *E. graminis* increased the apparent infection rate r_s of *Stagonospora* in the minimal interaction model and the rate parameter for infec-tious tissue in the extended interaction model. Altogether seven linked equations were developed for this analysis. Fits with non-zero interaction coefficients were obtained provided disease severity of one of the interacting pathogens was not too low. Weber (1996) assumed an inhibition of *S. nodorum* on *Erysiphe graminis* and an enhanced growth of *S. nodorum* favoured by the powdery mildew. With logis-tic growth of the curves, the sensitivity of parameters for the model most appro-priate for field and greenhouse data varied greatly when both pathogens were linked in the equation via their density regulation terms:

$$dy_m/dt = \max(0, r_m \cdot y_m(1 - y_m/K_m - \alpha \cdot y_s)$$

$$dy_s/dt = r_s \cdot y_s \cdot (1 - y_s/K_s + \beta - y_s)$$

where y_m and y_s stand for disease severity, and K_m and K_s are parameters for the maximum severity levels of *E. graminis* and *S. nodorum* respectively, without mutual interaction. The interaction parameters α and β (both >0) that denote the density inhibiting effect of *E. graminis* and *S. nodorum* respectively are restricted to positive values only, which hold for all models used by Weber (1996). In the above equations, leaf blotch is modelled as a competitor for infection sites. The ef-fect of mildew is included as a delay factor regulating the density of leaf blotch by making the interaction term αy_s positive. The term max(0) prevents the decline of the mildew growth rate as leaf blotch does not overgrow colonies of mildew.

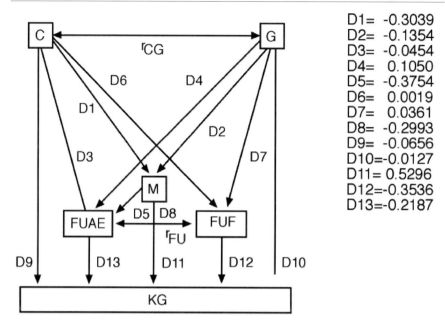

Fig. 4.14. Path correlation analysis for the effect of various constraints on yield of wheat (kernel weight KG). C = *Pseudocercosporella herpotrichoides*, G = *Gaeumannomyces graminis*, r_{CG} = correlation coefficient between C and G, SE = *Stagonospora nodorum*, M = "Morphology" a combination of various host parameters independent of disease (Table 3.5). The values of the path coefficients D stand for the positive or negative effects among variables measured in the field experiments. (from Jörg 1987)

The differential of *E. graminis* was limited to non-negative values in order to avoid a decrease of disease severity which might have resulted in an apparent "overgrowing" of *E. graminis* by *S. nodorum*. In Fig. 4.15, parameter sensitivities for $y_s(0)$ and r_s (both for *S. nodorum*) were compared. The equilibrium severity was higher when *Stagonospora* severity was lower at time t_m for the onset of the competing powdery mildew (m in the above equation). Thus, epidemics of *S. nodorum* with a slow initial development, i.e. low $y_s(0)$ or low infection rate r_s, can reach higher severities compared to those which start fast (high y_o, high r; Fig. 4.15). When compared to logistic growth, parameter sensitivity for the most appropriate model for the field data – when both pathogens are linked via density regulation terms – differs greatly (Weber 1996).

In a further model, Weber (1996) assumed the powdery mildew does not change the K_s of the leaf blotch.

$$dy_m/dt=\max\{0; r_m(1-\alpha\,y_s)y_m(1-y_m/[K_m(1-\alpha_s y_s)])\}$$

$$dy_s/dt=r_s(1+\beta y_m)y_s(1-y_s/K_s)$$

This model with the terms as above was also fitted to field data of disease progress in 1991. The following parameter values were obtained: r_m=0.29 and r_s=0.14, both per day, Km=0.66 and Ks=0.40, α=3.29 and β=57.83. This means, if disease incidence of *S. nodorum* increases by 1%, the rate r_m and the maximum disease intensity K_m of *E. graminis* are reduced by 3.29%. On the other hand, an increase in mildew by 1% results in an increase in leaf blotch by 57.83%, without a change in its maximum disease intensity. When both diseases occur together, then (in this case) the maximum attainable disease incidence of the powdery mildew is reduced, whereas the same parameter remains unchanged for leaf blotch though this level is reached much earlier.

Following Weber's approach, Hau (pers. comm.), assuming that in an interaction rate and capacity parameters are affected in the same direction, but to a different extent, proposed the general interaction model

$$dy_1/dt = r_1(1-\alpha_{12}y_2/K1)y_1(1-y_1)/[K_1(1-\beta_{12}y_2/K_1)]$$

$$dy_2/dt = r_2(1-\alpha_{21}y_1/K_2)y_2(1-y_2)/[K_2(1-\beta_{21}y_1/K_2)]$$

In these equations, the coefficients α_{ij} and β_{ij} stand for the mutual effects on the rates and maximum disease intensities. They are more flexible than the original Lotka-Volterra equations, but apparently have not been tested to explain the dynamics of interacting diseases and their epidemics.

Recently, comparisons on pest constraints in rice under various agricultural conditions have been published with a characterisation of injury profiles of all relevant pests in relation to the production situation in some countries of tropical Asia (Savary et al. 2000a). Pinnschmidt et al. (1995a) developed methods to identify relevant parameters and how to collect data

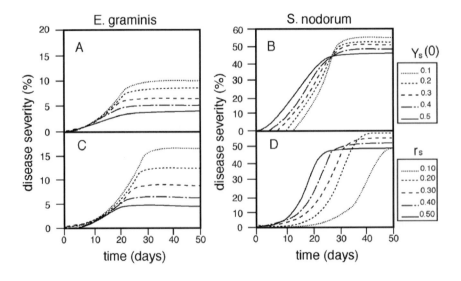

Fig. 4.15. Sensitivity of initial disease severity $y_s(0)$ (**A, B**), and infection rate r_s in (**C, D**) *Stagonospora nodorum* in the interaction model (see below). Standard parameter values were initial severity 0.2% and infection rate 0.2 for both pathogens, carrying capacity/interaction coefficient of 20%/4 and 40%/2 for *Erysiphe graminis* f. sp. *tritici* and *S. nodorum* respectively. (Weber 1996)

5 Comparison of Temporal Aspects
of Epidemics: The Disease Progress Curves

Disease progress curves (DPCs) as graphs of the temporal dynamics of epidemics emanate from plotting absolute or relative values of disease intensities y over time t on the x-axis. Naturally, disease progress is non-linear, though it may be adequately described by linear statistical methods, if necessary after appropriate transformation. Epidemics, as strongly influenced by variable or heterogeneous external factors are, by nature, rather stochastic. This, however, is not taken into account much in plant disease epidemiology when measuring and analysing epidemics. The reason is the possibility that large and repetitive measurements yield larger data sets (than in human and animal epidemiology), embracing much of the imminent variability.

Temporal dynamics of plant disease epidemics are measured by disease intensities y(t) which can be either plotted as summation curves against y, or rate curves against dy/dt on the y-axis. Disease assessment, however, is prone to varying degrees of uncertainties, mainly from inadequate sampling methods and errors in disease estimates (Hau et al. 1989). The curves obtained may be unilateral, bilateral (but different from rate curves), or bi- and multi-modal (Fig. 5.1) which express the temporal behaviour of the epidemic as a system resulting from interactions of systems structures stimulated by external factors within the disease square. Rate curves are usually bilateral (see Fig. 5.2). General aspects of DPCs (Fig. 5.1) have already been compared in some detail (e.g. Kranz 1968c, 1974a, 1978; Waggoner 1986; Campbell and Madden 1990; Gilligan 1990b, 1994; Campbell 1998). Campbell and Benson (1994a) compared disease progress curves of soilborne diseases. There is no evidence that progress curves differ essentially among pathosystems caused by the various groups of pathogens, and abiotic diseases, although progress of abiotic diseases tends to be polyetic.

Apart from the patterns of disease progress curves there are a number of descriptive terms to compare epidemics (Gäumann 1951). Explosive epidemics increase rapidly to a culmination point and then decline (i.e. the classic epidemic, curve c in Fig. 5.1). Tardive epidemics, like root rots of tree crops, progress slowly and reach rather late peaks, plateaux or an asymptote, if any. Annual epidemics are commonly studied in epidemiology as well as polyetic epidemics progress over several seasons or years. Pandemics spread over large distances or continents and often cause severe disease. There is a threat of great damage when such a pathogen becomes established in a new area. Pandemics usually subside in strength, once the susceptible host genotypes have been eliminated. For crop loss assessment, James (1974) distinguished short and long epidemics, as well as early and late predictive disease. Patterns of epidemics are also compared by reference to some famous or well-documented or described epidemics. Examples of such historical epidemics are the "Heines VII-yellow wheat rust epidemic" (Zadoks 1961) and the northern and southern Japan types of rice blast epidemics (Kato 1974).

Chapter 5 deals with the comparison of temporal aspects of epidemics and some determining factors, starting with endemicity as a special case (Sect. 5.1). Thereafter, structural elements of disease progress curves will be treated (Sect. 5.2), followed by unilateral (Sect. 5.3), bilateral und multilateral epidemics (Sect. 5.4), and finally, disease progress in soilborne diseases (Sect. 5.5).

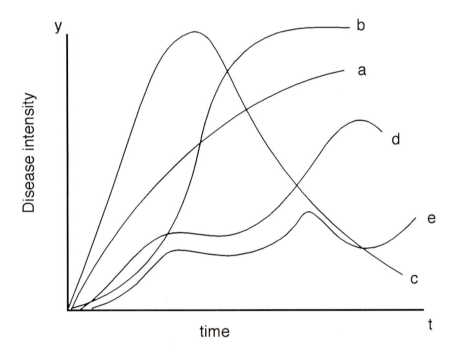

Fig. 5.1. Various shapes of disease progress curves: *a* unilateral, *b* sigmoidal, *c* bilateral, *d* bimodal, *e* multi-modal. (Kranz 1996)

5.1 Endemicity

The current definition of epidemics in plant pathology with disease intensity y in the range of (0, y≤1) or (0, 1) includes endemics. Gäumann (1951) defined endemics in a geobotanic sense as diseases caused by pathogens that are "at home" in an area. Similarly, Van der Plank (1975) defines a disease as endemic when the pathogen has coexisted with its host in the same habitat long enough to be adapted to the environment, but not a pathogen necessarily native to the area. For an endemicity sensu Van der Plank (1975), it is essential that an endemic pathosystem exists under the prevailing conditions in a state of almost timeless balance and co-existence with the host. This implies an average progeny/parent ratio $\alpha = (iR) \approx 1$, as long as tissue available for infection is unlimited. Van der Plank quotes virus diseases of (forest) trees as an example. Obviously, in endemicities, the population resistance of the host plants is adequate to keep the disease at the threshold level $iR \approx 1$. This ensures the survival of the pathogen at low disease intensity without causing obvious harm to the host plant. A slow buildup with an infection rate near zero is the consequence. However, flare-ups and conspicuous outbreaks can occur. Epidemics and endemics of the pathosystem then form a continuum (Gäumann 1951). Endemic diseases may even turn into severe epidemics in newly encountered host plants. For instance, swollen shoot virus disease of cacao (CSSV) in West Africa has been transmitted from local trees, where the virus is inconspicuous, to cacao when this crop was brought to the region (Thresh et al. 1988). Endemic diseases usually are polyetic with a rather slow temporal and spatial increase in disease intensity.

For Onstad and Kornkven (1992) endemicity is the "persistence or constant presence of a pathogen in an ecologically proper spatial unit over many generations". This implies a progeny/parent ratio of about 1. By means of analytical models of four linked differential equations, they identified an "ecologically valid spatial unit" as determining host factors for endemicity: (1) absolute density of susceptible hosts and (2) growth and relative density of susceptible leaflets. Less dense host stands decreased persistence, whereas continuous growth of susceptible host tissue increased the probability of persistence, according to their models. For pathogen and disease this is due to: (1) average magnitude and heterogeneity of iR and (2) the duration of the pathogen's generation time or infection cycle. With decreasing iR the pathogen in their simulations was less likely to persist for a given number of generations. Otherwise, the decrease was more likely with longer infection cycles. Jeger and van den Bosch (1994b) analysed the Onstad and Kornkven system of equations and agreed with them. However, Jeger and van den Bosch suggested some modifications of the host-related parts of the models and added some more features, including the invasion argument.

"Crowd diseases" are a special case of endemicity (Van der Plank 1948). Pathogens that cause such diseases do not persist in the soil and do not spread quickly or to great distances in any considerable amount because of low rates and steep gradients. Cacao swollen shoot virus (CSSV) is considered a prime example of crowd diseases as long as it occurs on natural host trees in West Africa (Thresh et al. 1988). It became widespread once cacao was planted in the proximity to natural hosts of the virus. In contrast to crowd disease, a "vagile disease" like the African cassava mosaic spreads quickly by means of its vector *Bemisia tabaci* (Thresh 1991) and cannot be controlled by isolation or sanitation, as is possible with CSSV. Thresh (1991) regards another virus disease, the maize streak virus, as a crowd disease in some circumstances or certain times of the year and as vagile at other times.

5.2 Structural Elements of Disease Progress Curves

Components or elements related to phases of disease progress curves integrate biologically relevant effects on epidemics as a system. Kranz (1978) and Vanderplank (1982), with different concepts, attempted an "anatomy of epidemics". Kranz (1974a, 1978) emphasised the curve structures, whereas Vanderplank (1982) stressed underlying processes such as progeny/parent ratio, the role of latent periods, dwindling inoculum, threshold conditions, and the internal checks and balances. Both approaches to an anatomy of epidemics are complementary and will be dealt with in this section. A more mathematical comparison of DPCs has been made by Gilligan (1990b)

5.2.1 Criteria for Disease Progress Curves

Elements of structures of an epidemic are compared by their state variables $y(t)$, e.g. disease intensity, their functions $f[y(t)]$, for instance, inoculum produced, loss inflicted by a given level of disease intensity at time t, and rates. State variables express the state of an epidemic at any time t or point d in space. Behind state and rate variables there are internal structures, like α, β and γ in Fig. 2.1, which operate as checks and balances through recurrences, delays, thresholds and limits, and factors that affect them. Structural elements could be subsystems of a defined system, like the components of the infection chain (Sect. 4.3). If the role and function of structural elements in the interaction are known, they can be relevant criteria to compare internal structures of epidemics. These may reveal whether spore production, spore dispersal or survival of the inoculum have the greater impact on disease progress, are more sensitive to compensation (Sect. 4.4.2) or are more easily controlled. Such comparisons could be made with descriptive or measurement terms tested statistically.

Although each element is indispensable to the total structure, its quantitative impact or weight varies, conditioned by its previous state and by various external factors. When the system "epidemic" is dissected for its apparent and intrinsic structures, one should keep in mind that the behaviour of single elements does not represent the behaviour of the entire structure. The whole is more than the sum of its elements (p. 3). The question invariably arises of how far a cutting-up of an entire epidemics should go because in practice not every component and all its interactions are known, measurable or relevant. Therefore, elements that affect the "core dynamics" (Patten 1971) rather would suffice for a comparative purpose.

Interactions between state variables within the disease square and their external rate-determining factors result in various patterns of epidemic behaviour – the disease progress. As a consequence, kind, strength and flexibility of the interactions within the disease square contribute to the behaviour of an epidemic. Cracraft (1981; cited by Wenzel 1992) brings it to the point "... that there may be one or more functions for a structure which does not imply an associated stereotypic behaviour pattern for each function; it is the behaviour pattern, then, and not the function, which constitutes a systematic character." A "systematic character" in this chapter is the disease progress curve as the graph of the temporal dynamics in epidemics, which for Thresh (1974a) meant each system is effective in different ways with a corresponding diversity in epidemiology.

Elements as criteria of complete, bilateral disease progress curves are described in Fig. 5.2. In addition, the following are descriptive and measurement terms of the DPC after Kranz (1974a).

From the schematic bilateral disease progress curve in Fig. 5.2, the following variables with their parameters can be derived (time in days):

1. t_B: time of the beginning of host vegetation or refoliation.
2. t_0: time when primary outbreak $y_0 = t(y_0)$ was observed.
3. t_1: time when after the inflection point ip of the progressive leg the disease progress turns into the asymptote or plateau near y_{max}.
4. t_2: time when disease progress starts declining. t_2 is equal to t_F when the epidemic, for instance, ends at harvest.
5. t_F: time when the epidemic ends (natural or harvest).
6. $d_0 = t_B - t_0$: time elapsing from t_B to t_0, the disease-free period.
7. $d_p = t_1 - t_0$: duration of the progressive leg.
8. $d_{max} = t_2 - t_1$: time period during which the disease intensity is near to y_{max}. If disease progression is asymptotic, t_2 is equal to t_F.
9. $d_D = t_F - t_2$: duration of the degressive leg (for bilateral curves only).
10. $d_T = t_F - t_0$: total duration of a naturally completed epidemic.

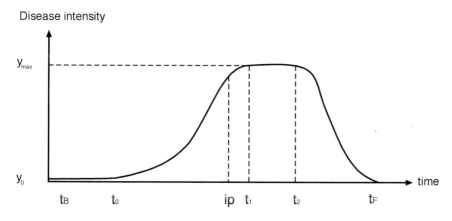

Fig. 5.2. The elements of a disease progress curve as graph of an epidemic. For explanations, see text. (Kranz 1974a, modified)

11. $d_N=t_o-t_F$: duration of the time interval between the disappearance of last and reappearance of the following epidemic.
12. y_o: intensity of primary disease outbreak ($0<y\leq1$).
13. y_{max}: the maximum disease intensity ($y_{max}\leq1$), usually the disease intensity of the peak, plateau or asymptote. The y_{max} may correspond to K, the carrying capacity of a crop for disease intensity, or the asymptote (Vanderplank 1982).
14. y_F: final disease intensity at t_F, the end of the epidemic. It is possible that y_F is equal to 0 or equal to y_{max}.
15. AUDPC: area under disease progress curve from t_o to t_F.

The infection rate will be treated in Section 5.2.3.

Damage caused by disease may by expressed as yield loss (see Sect. 7.1), defoliation or proportion of host plants killed (e.g. fruit rots) and measured by parameter values of disease incidence or severity. This criterion is included as term BF in Table 5.1.

Obviously, small changes in parameter values, for instance in the initial conditions or small variations in environmental factors, can markedly change the effects of one of the above parameters on others in the system. This relative importance among structural elements in Fig. 5.2 is largely determined by their degree of variability and intercorrelation (Kranz 1968b). This can be elucidated by means of principal axis Factor Analysis (PAFA) with rotation. The PAFA was employed to reveal intercorrelation among 13 elements of 40 different pathosystems investi-

Table 5.1. Factor analysis of curve elements, factor numbers and their loads. (Adapted from Kranz 1968c)

Curve element		Factor number and load					
		1	2	3	4	5	6
d_o	Delay of start epidemic	0.03	−0.24	−0.29	−0.17	0.81	−0.12
d_T	Duration of epidemic	−0.06	0.48	0.53	0.65	−0.19	0.01
d_p	Duration of progressive leg	−0.05	−0.18	0.94	0.04	−0.10	−0.21
d_{max}	Duration of asymptote	−0.01	0.97	−0.14	0.05	−0.10	0.10
d_D	Duration of degressive leg	0	−0.01	−0.01	0.99	0.06	0.07
y_o	Initial disease intensity	0.76	0.12	−0.14	−0.21	0.07	0.24
y_{max}	Maximum disease intensity	0.87	−0.18	−0.05	0.05	−0.13	−0.19
r^a	Infection rate	0.11	0.09	−0.19	0.06	−0.11	0.93
AUDPC		0.73	0.12	0.33	0.23	−0.08	0.38
BF^b		0.57	−0.16	−0.26	−0.05	−0.64	0.05
Eigenvalues		2.20	1.33	1.50	1.54	1.16	1.19
Share of variance explained^c (%)		24.6	15.0	16.8	17.3	13.0	13.3

For definitions of the curve elements as variables see Fig. 5.2; the h_i^2 values per line range from 0.73 (for y_o) to 0.98 (for d_D); the sum of the eigenvalues is 8.92.
[a] The apparent infection rate is not a curve element in Fig. 5.2.
[b] BF=consequences of attack by disease in six nominal classes (Kranz 1968a).
[c] Calculated as (eigenvalues$_{ij}$/sum h_i^2) × 100.

gated for 2 years (p. 128). Six factors (1–6) were extracted from the rotated correlation matrix for the ten quantitative elements (Table 5.1). Elements in a factor
with bold letters showed the highest intercorrelation in this factor. However, the
resulting factor loads in the varimax Factor Analysis may be difficult to interpret
(Mardia et al. 1979) unless the variable has a high load (e.g. >0.35) on at least one
factor. The factor loads in Table 5.1 are defined by the programme in the order of
the variance explained by their eigenvalues.

The more important variables or curve elements in Table 5.1 are in the factor 1
with the highest share of explained variance, but even factors 5 and 6 are plausible
(see below). Factor 1 proves a high and positive intercorrelation between initial
disease intensity y_0 and maximum disease intensity y_{max}. These factors, in turn,
tie in with the AUDPC (Van der Plank 1963), and BF for "consequences of the
epidemic" or damage to the host plants (Kranz 1968b). Therefore, a high primary
infection very often results in a high final level of disease, a large area under the
disease progress curve and a high proportion of plants killed. These all have been
corroborated, for instance, by Gassert (1978; Table 5.4), and Plaut and Berger
(1981).

The factors 2, 3, and 4 all have high loads in the duration of the epidemic and
consequently are tied to parts of the disease progress curve. In factor 2 it is the duration of the plateau or asymptote, and in factor 3, the duration of the progressive
leg of the curve. Factor 4 finally has a high load for the degressive leg of a bilateral curve. Surprisingly, none of these factors shows a load >0.35 for the AUDPC,
although factor 3 comes close to it (0.33).

Though factor 5 has a low share of variance explained in this study, it indicates
a negative correlation between the lag between "date of host emergence to first
visible disease", or delay of disease onset, and "consequences of the epidemic."
Early disease is thus more likely to result in severe disease, and vice versa. Hamilton and Stakman (1967) observed wheat stem rust for 42 years in the Mississippi
Basin and they related y_{max} to the date of the first rust appearance. The earlier the
first incidence, the higher y_{max} tended to be. The delay "start of the epidemic" d_0
is of prognostic value for the time of onset of spraying, for the "negative progno-
sis" for *Phytophthora infestans* on potato (Schrödter and Ullrich 1965; Ullrich and
Schrödter 1966) and *Leveilulla taurica* on tomatoes and pepper (Palti 1971). In
spite of its fairly high consistency in Table 5.1, t_0 may be somewhat variable in
the field (Kranz 1968c). The y_0 often remains too low to be perceived and assessed for the onset of the epidemic, thus the onset appears delayed.

Finally, factor 6 is of interest as it conforms with Van der Plank's (1963) find-
ing of a close correlation between the apparent infection rate and the AUDPC in
spite of a rather large variability of r causing the inferior rank of this factor (see
Sect. 5.2.3). The AUDPC, according to Waggoner (1986), summarises the seasonal course of an epidemic better than the rate r. In the context of the study
(Kranz 1968a), the infection rate and the AUDPC rather appeared to be functions
of the environment.

With the same programme for Factor Analysis, Campbell et al. (1980) performed a comparison of eight elements of disease progress of bean root rot epidemics. They analysed slightly different variables (Table 5.2) from the ones in
Table 5.1.

Table 5.2. Principal axis Factor Analysis with varimax rotation utilising eight curve elements to characterise 100 naturally occurring bean root rot epidemics caused by *Rhizoctonia solani* and *Fusarium solani* f. sp. *phaseoli*. (Campbell et al. 1980)

Variable	Factors with load			
	1	2	3	4
Weibull[a] scale parameter b	0.84	−0.29	0.05	0.07
Weibull shape parameter c	0.93	−0.05	0.06	−0.03
AUDPC	0.07	0.96	−0.11	0.20
Final disease severity y_f[b]	0.64	0.65	−0.01	0.35
FDR[c]	−0.85	−0.03	−0.07	−0.33
Regression quadratic coeff.[d]	0.84	0.24	0.02	0.38
Disease incidence (%)	00.14	0.25	−0.07	0.90
Time of disease onset	0.04	−0.09	0.99	−0.06
Variance	3.4	1.6	1.0	1.2
Explained variance (%)	42.3	20.0	12.5	15.0

[a]The Weibull growth function (see Table 5.13).
[b]After 50 days.
[c]FDR, the first-difference linear regression coefficient was computed in a regression analysis to describe the disease progress curve which in addition to the common regression coefficient b included a quadratic b coefficient and an error term e.
[d]The quadratic regression coefficient from the FDR.

Campbell et al. (1980) interpret the first and main factor as an overall description of disease progression, including curve shape c, scale, rate (as b) and final severity y_F, among the parameters in Table 5.2. The shape parameter c describes the nature of an epidemic, the greater the value of c, the greater y_f. The FDR coefficient indicates that a linear term in the equation tends to limit the rate of disease increase relative to the effects of the quadratic coefficient. The consequence is a negative intercorrelation between y_f and FDR, and a positive one with the quadratic coefficient. The scale parameter b is inversely proportional to the rate of disease increase with a negative intercorrelation between the quadratic b coefficient and y_f. In factor 2, AUDPC and final disease severity y_f are positively intercorrelated, as can also be seen in Table 5.1. Factor 3 is interpreted by Campbell et al. (1980) as the epidemic location factor which "identifies the uniqueness of the time of disease onset with respect to their curve elements". For the two root diseases, the third factor (location of epidemic) did not show the strong relationship between the time of onset of disease t_o and y_{max}. The authors conclude that breeding for resistance to delay the onset of the epidemic would not be useful as it does not delineate well with other curve elements. Factor 4 just confirmed the relationship between disease incidence and severity. This correlation is valid only up to a certain level of disease incidence, for instance, up to 70% (Koch and Hau 1980).

Waggoner (1986) defined the logistic curve by the initial number of infections y_o, the rate of disease progress r and the maximum disease intensity K (y_{max}). The curve ideally is symmetrical around the inflection point K/2 as with logistic curves, the skewness of which is zero. He added skewness as a further criterion; real progress curves are not exactly symmetrical, but either negatively or positively skewed. The skewness S of the curves, therefore, is seen by Waggoner as an important curve characteristic. The calculation of S requires estimates of the variance and the third moment. The skewness is dimensionless with a standard error $SE=[6/(\text{number in sample})]^{1/2}$. In Table 5.3 the skewness of disease progress in three pathosystems is compared, as calculated by Waggoner (1986, Eq. 5).

Table 5.3. Comparison for skewness S as a feature of disease progress curves representing epidemics of *Phytophthora infestans* on potatoes, *Cercospora apii* on celery, and *Uromyces appendiculatus* on beans with standard error SE and coefficients of determination CD of regressions with logits and gompits G. (Adapted from Waggoner 1986)

Parameters	*P. infestans*		*C. apii*		*U.appendiculatus*
	Unsprayed	Sprayed	EES	FLA[a]	
Skew, S	1.5	0.52	0.45	–0.22	–1.51
SE	0.20	0.20	0.37	0.37	0.07
Regressions					
CD (%) logit	94	89	96	99	82
CD (%) gompit	98	95	96	90	82

[a]Two cultivars, EES 1624 and FLA 683.

For potato late blight, skewness of the disease progress curves is positive (shifted to the right) and partly negative (increasing early in the epidemic) for celery blight and bean rust (increasing late in the epidemic). The coefficients of determination for the regressions in Table 5.3 differ among the pathosystems. For late blight the gompits show a higher coefficient of determination (CD) than the logits when sprayed, and one of the curves of celery also differed.

Ngugi et al. (2000) compared the epidemics of the anthracnosis (*Colletotrichum sublineolum*) and leaf blight (*Exserohilum turcicum*) of sorghum in Kenya by means of a three-parameter, non-linear logistic model $Y_t=\gamma/1+\exp[-\beta(t-\eta)]+\varepsilon_t$, where γ (%) stands for the final disease severity denoted by the upper asymptote, β is the rate for disease progress per day, and η in days after emergence. From this equation they derived the following parameters to describe disease progress curves relative to growth stages: (1) V_{95} by inserting 95 (days) for t_1 in the model, (2) $t_2=\{-\log[(\gamma/2)-1]/\beta\}+\eta$ to assess the time needed to reach 2% disease severity as the start of the epidemic at $t_y=2$, a time at which disease could be first observed and (3) an absolute rate parameter θ ($\beta \times \gamma$) was estimated to facilitate comparison of disease progress curves having different asymptotes at the milky stage 95 days after emergence, y(t=95). For results of these comparisons see Section 7.3.1. The terms used by the authors are essentially the same as in Tables 5.1 and 5.2.

Navas-Cortés et al. (1998) compared various factors that affected the development of *Fusarium* wilt of chickpea (pp. 170–171). They used five curve elements to characterise DPCs obtained in field experiments: viz. (1) DII = disease intensity index at the final date of disease assessment (i.e. y_F); (2) SAUDPC = area under disease intensity progress curve estimated by the trapezoidal integration method standardised by duration time in days; (3) DII(t_{ip}) = disease intensity index at the inflection point of the DPC calculated by the estimates of parameters for the Bertalanffy-Richards function; (4) t_{ip} = time in days to reach the point of inflection; and (5) $t_{0.05}$ time in days to reach initial symptoms, estimated as the number of days to reach DII = 0.05. The conclusion of Navas-Cortés et al. (1998) was that in southern Spain shifting the sowing date of chickpeas from early spring to early winter can slow down the *Fusarium* wilt epidemic, delay the onset of the epidemic and minimise final disease incidence (y_F). This may, however, be somewhat influenced by the susceptibility of the cultivar and the virulence and density of the inoculum in the soil.

Vanderplank (1982) in his anatomy of epidemics treated essential parameters that affect processes of epidemics. These are: (1) the initial disease intensity y_o; (2) the progeny/parent ratio α as net reproduction rate per generation; (3) the latent period p; and (4) the apparent infection rate r. For the progeny/parent ratio $\alpha=iR$, the duration of the infectious period i is the unit of time (Vanderplank 1982). He further dealt with dwindling inoculum and the threshold conditions for an epidemic in the context of α. The level of the asymptote L (Sect. 5.2.4) could be another criterion for comparison. Vanderplank's (1982) concept of an "anatomy of epidemics" also includes the "internal checks and balances" in epidemics, i.e. structural elements like limits, thresholds, delays etc. (Kranz 1980 p. 7), and their functions in the output–input relationships between two successive elements (e.g. the output of disease intensity y(t) and its input for y(t+1), both being state variables for t. These interactions may be modified by the impact of external factors, e.g. course of weather events, control practices and genetics of host and pathogen.

5.2.2 Initial and Maximum Disease Intensities

Elements which affect the start of epidemics are Q (see Table 3.11), the surviving quantity of inoculum, the primary disease y_o, and t_o, the time at which primary disease occurs. Larger Q under favourable infection conditions appear to be consistently correlated with higher initial disease intensity y_o, earlier disease outbreak t_o and higher y_{max} (Kranz 1968c). These conditions more strictly apply to the first infection cycle only. Removal of mummified coffee berries, which harbour the bulk of overseasoning inoculum of *C. coffeanum*, reduced the primary y_o to 1 versus 7 in the plots with no removal of mummies. Thereafter, the effect of the delay Δt very much depends on the rates during the ensuing epidemic.

There is a clear effect of sanitation by the removal of mummified coffee berries from the trees. The faster rate r at low y_o of CBD in the plot with inoculum removed, however, reduced the initial ratio of lesions of 1:7 to a ratio of 1:2 between both treatments. As the rates during the following weeks remained practically the same in "removed" and "not removed", this ratio did not change until y_{max} was reached; still a beneficial, though not proportional sanitation effect remained.

Table 5.4. The effect of reduced Q by removing dried-up and overseasoning coffee berries ("mummies") from coffee trees which harbour most inoculum of *Colletotrichum coffeanum* for y_0 at $t(y_0)$, i.e. 13th week, of coffee berry disease (% incidence[a]) and further disease progress $y(t)$ and apparent infection rates r. (Gassert 1978)

Week after flowering	No mummies		With mummies	
	y(t)	r	y(t)	r
13	0.09		0.62	
		0.68		0.50
14	9.24		17.10	
		0.10		0.11
15	17.11		30.35	
		0.042		0.035
16[b]	21.80		38.90	

[a]Incidence in percent are based on lesion counts.
[b]No further increase in incidence, as berries had turned resistant.

The greater or the earlier y_0, the higher y_{max} tends to be rather common among pathosystems, whilst a late start of the epidemic rather lowers the y_{max}. For northern and southern corn leaf blights (*Exserohilum turcicum* and *Bipolaris maydis*, respectively), y_0 in the field was the most obvious factor in determining final disease intensity y_{max} (Berger 1973, 1974; Sumner and Littrell 1974). Across 59 fungal pathosystems in wild plant populations, early t_0 and high y_0 were positively correlated with high values of y_{max} (Kranz 1968b), larger area under disease progress curves and a higher frequency of plants killed. The earlier the first incidence of wheat stem rust (*Puccinia graminis*) was in the Mississippi Basin, the higher y_{max} used to be (Hamilton and Stakman 1967). This is a phenomenon which is also known for virus diseases (Thresh 1974b). Juvenile resistance retards epidemics and disease progress (e.g. as slow rusting). A delay in the onset of epidemics of black rot of groundnuts (*Cylindrocladium crotolariae*) was more important than the rate of disease progress in moderately resistant and susceptible genotypes (Culbreath et al. 1991). The area under the disease progress curve (AUDPC) and the yield loss were greater with higher initial blight incidence (y_0) of *Septoria apiicola* on celery at transplanting (Mudita and Kushalappa 1993).

The distribution of primary inoculum in a field evidently also has its impact on the onset and further behaviour of epidemics and on rates (Sect. 5.2.3). Inoculum of seedborne pathogens (Baker and Smith 1966; McGee 1995), or infected and transplanted seedlings (Berger 1973) tends to be more often randomly distributed in fields than pathogens dispersed by other means. This creates a maximum opportunity for the infection of young neighbouring plants, particularly through matching virulent races to susceptible cultivars. Hence, the number of dispersal and infection cycles required to cover a given crop area may be less with seedborne diseases than with diseases that develop from more localised, but equally strong primary foci. For this reason, other things being equal, epidemics of seedborne pathogens will develop at more rapid rates (Sect. 5.5).

In the example in Table 5.4 with coffee as a non-deciduous crop, autoinfection prevails throughout the vegetation. However, in annual or deciduous perennial crops both allo- and autoinfections occur. According to Robinson (1976), allo-infections are predominant during what he calls the exodemic. The exodemic is the early phase of an epidemic when the canopies of the host plants are still open. Pathotypes in airborne inoculum have an equal chance to land on the plants and infect. The esodemic phase of an epidemic prevails after the canopy closes and when susceptible host tissue is available. The majority of inoculum is then pro-duced within the field. Where all plants have the same resistance genotype auto-infections are predominant. An example of the effect that these two phases of an epidemic have on the selection of pathotypes is in Table 5.5. The pathotype (race) with the virulence genes Vg-Va6-Vv as compatible with the host resistance gene Mlv increased its frequency due to the selection pressure of the Mlv resistance gene. From 2% in the exodemic during the esodemic it reached 35% in one season and topped the other identified pathogen genotypes.

Table 5.5. Selection of races during the exodemic and esodemic in samples of curve Varunda with the R-genes Mlg and Mlv. (Eckhardt, pers. comm.)

Genotype (race) of pathogen	Proportion (%) of races during			
	Exodemic	Esodemic		
	31.5.	8.6.	14.6.	27.6.
Vg-Va6-Va12	43	24	46	27
Vg-Va6-Vv	2	20	22	35
Vg-Va6	35	28	14	3
Rest of genotypes	20	28	15	33

There is an indication that critical disease intensities $y(t)$ may exist, which, once surpassed, give way to a general outbreak and make control more difficult. Such start-up thresholds differ greatly from one disease to another. In studies on potato late blight over 5 years, Hirst and Stedman (1960) found that the frequency of stem lesions must be of the order of 0.52–0.79% before the epidemic started to "take off." The computer-based negative forecast PHYTPROG (Schrödter and Ullrich 1965) for potato late blight (*Phytophthora infestans*) predicts the 1% se-verity threshold as the point to commence spraying in number of days after emer-gence of the crop. Rice blast epidemics in some areas of Japan show two distinct phases of progress. During the first phase the disease intensity gradually reaches the 5% level, thereafter the infection rate increases steeply (Kato 1974). With cof-fee berry disease (*Colletotrichum coffeanum*) this happened after 3–5% of the ber-ries had been infected (Steiner 1973; Fig. 5.4). *Cercospora apii* on celery took off when a 2% disease intensity was surpassed (Berger 1975). No such threshold to initiate fungicide application could be established for *Septoria apiicola* on celery (Mudita and Kushalappa 1993). So far, however, it is not clear whether a take-off level may be of general application. Comparative posterior analyses could possi-bly answer this query.

5.2.3 Comparison of Infection Rates

Rates describe the speed of change of states, e.g. disease intensity, in time. In epidemiology, they also are the derivatives of differential and difference equations used as growth functions (Table 5.13). The absolute rate dy/dt (Fig. 5.3) describes the instantaneous change of a state variable dy, e.g. for disease intensity, during an infinitesimal time interval dt. The relative rates r derived from the absolute rates are parameters for the speed of response in a system to a stimulus. Thus, these rates describe the dynamics as a discrete or continuous sequence of changing states responding to external (e.g. weather, treatments) or internal factors (degree of resistance etc.). Rates are thus mere speedometers (Van der Plank 1975) of changes in states over time (and distance) and are meaningful only in the context for which they have been calculated. The more commonly used rates are the exponential (logarithmic sensu Van der Plank $r_l=[\ln(y_2)-\ln(y_1)]/(t_2-t_1)$, and the apparent or logistic infection rate $r=[\ln(y_2/(1-y_2))-\ln(y_1/(1-y_1))]/(t_2-t_1)$ proposed by Van der Plank (1963): with disease intensity y_1 and y_2 at dates of measurement, t_1 and t_2. For other growth functions, e.g. the Gompertz function, rates are defined in Table 5.13. Apparently, no essential differences exist among rates of epidemics caused by viruses, bacteria or fungi. However, among taxa and pathosystems, differences in rates are obvious sometimes with wide ranges caused by affecting factors (Table 5.6). More mathematical details on rates derived from growth functions can be found in Table 5.13.

Rates vary because of their sensitivity to affecting factors (Table 5.6), a fact also stressed by Waggoner (1986). A comparison of rates may, therefore, still be useful particularly for immediate comparisons of environmental or treatment effects and resistance to the epidemic behaviour. But it may limit the suitability of rates as criteria for comparison (Kranz 1968c) unless fairly natural classes of diseases can be delimited as, for instance, in Table 5.7. Nevertheless, rates are important parameters for growth functions, analytical and simulation models. For "strategic" computations and simulations it would, therefore, be useful to know the range r-values can have in a pathosystem and a possible frequency distribution of observed values.

Only five pathosystems in Table 5.6 had the same r-values in both the humid and the dry year. In 10 of the 30 pathosystems, rates were higher (+) while 15 had lower rates (–) in the dry year, particularly among *Alternaria*, *Cercospora* and *Oidium* spp. (Kranz 1968b). For half of the Uredinales, the r-values increased. There also seems to be an effect of the host plant: *Phyllachora digitariae* and *Aecidium justiciae* behaved in opposite ways on different hosts. Species of *Cercospora*, *Phyllachora* and *Puccinia* had mean r-values of 0.18, 0.16 and 0.12 respectively. In addition, the epidemics of *Phyllachora* spp. tended to cause rather frequent and rapid total breakdowns of their host populations. *Cercospora* spp. attacked their hosts earlier and over longer periods. Both *Phyllachora* spp. and *Cercospora* spp. also had a greater area under progress curves than did the epidemics of *Puccinia* spp.

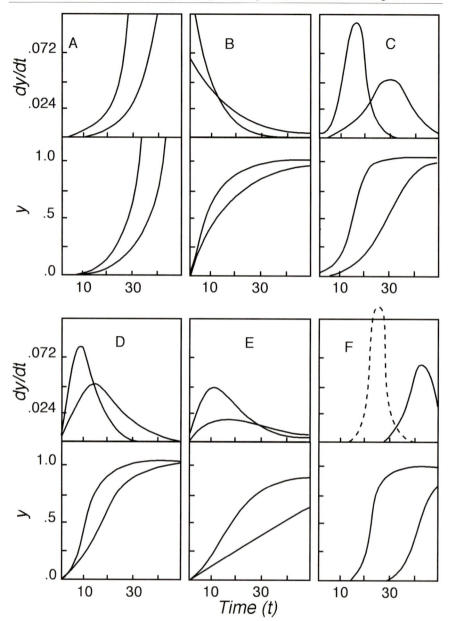

Fig. 5.3. Curves of absolute rate (*dy/dt* versus *t*) and disease intensity (*y* versus *t*) at two relative rates r for six growth functions used to describe disease progress: **A** exponential, **B** monomolecular, **C** logistic, **D** Gompertz, **E** log-logistic, and **F** Bertalanffy-Richards with m=2.7. (Campbell and Madden 1990, with permission)

Table 5.6. Apparent infection rates r of selected wild pathosystems studied for 2 years at the same sites in Guinea. (Kranz 1968b)

Pathosystem	r-Values in year[a] being	
	Humid	Dry
Ageratum conyzoides – Alternaria ziniae	0.19	0.10 –[b]
Bidens pilosa – Cercospora bidentis	0.30	0.06 –
Borreria verticillata – Puccinia lateritia	0.19	0.12 –
Borreria verticillata – Mycosphaerella sp.	0.05	0.10 +
Calopogonium muconoides – Cercospora canescens	0.06	0.11 +
Calopogonium muconoides – Corticium solani	0.08	0.08
Chasmopodium caudatum – Cercospora fusimaculans	0.25	0.20 –
Chasmopodium caudatum – Meliola panici	0.25	0.07 –
Cleredendron scandens – Cercospora volkamaria	0.19	0.03 –
Cleredendron scandens – Hemileia scholei	0.26	0.08 –
Commelina benghalensis – Kordyana celebensis	0.15	0.19 +
Crinum sanderianum – Aecidium crini	0.15	0.20 +
Crotolaria retusa – Uromyces decoratus	0.06	0.09 +
Crotolaria retusa – Alternaria tenuissima	0.13	0.07 –
Cyperus rotundus – Puccinia conclusa	0.15	0.30 +
Digitaria horizontalis – Phyllachora digitariae	0.15	0.08 –
Digitaria longiflorae – Puccinia levis	0.15	0.08 –
Digitaria longiflorae – Phyllachora digitariae	0.05	0.16 +
Dioscorea hirtiflora – Cercospora contraria	0.08	0.08
Dioscorea sp. – Cercospora carbonacea	0.08	0.15 +
Euphorbia hirta – Peronospora favargeri	0.02	0.03
Ipomoea involucrata – Cercospora timorensis	0.23	0.08 –
Justicia sp. – Aecidium justiciae	0.03	0.20 +
Oplismenus hirtellus – Plasmopara oplismeni	0.06	0.08
Pennisetum subangustum – Cercospora fusimaculans	0.13	0.07 –
P. subangustum – Phyllachora sphaerosperma	0.15	0.08 –
Phaylopsis falsisepala – Aecidium justiciae	0.26	0.07 –
Spigelia anthelmia – Oidium sp.	0.23	0.15 –
Spilanthus acmella – Oidium sp.	0.30	0.19 –
Tridax procumbens – Cercospora tridax – procumbentis	0.01	0.08 +

[a]Annual rainfall in humid year 2570 mm, and 1680 mm in the dry year.
[b]The – and + indicate whether r was lower or higher in the drier year.

Disease progress and rates may differ between herbaceous and woody plants. At least epidemics of virus diseases increased rapidly in herbaceous annual crops, but slowly in woody perennials despite the longevity and large canopies of the latter as catchment areas for vectors. Tristeza virus disease of citrus is an exception. Thresh (1983) explained the frequently low rates of virus diseases of woody host plants by the following facts: (1) woody plants are more difficult to infect compared to herbaceous annuals; (2) they are slow to be invaded and develop systemical symptoms; (3) woody plants tend to be tolerant to infection and frequently develop inconspicuous symptoms that are associated with a low availability of virus to vectors; (4) woody perennials are usually grown at wide spacing so that roots and branches of adjacent plants are seldom in contact which impedes dissemination. However, disease progress curves of virus diseases are similar to

Table 5.7. Proposed classes for apparent infection rates r[a] defined by medians and class limits.

Class No.	Range limits	Median
	r	r
I	0.01–0.05	0.03
II	0.06–0.10	0.08
III	0.11–0.16	0.14
IV	0.17–0.25	0.20
V	0.26–0.30	0.28
VI	0.31–1.00	0.40

This is a tentative example deduced from own experiments (Kranz 1968c). Other limits and medians for the infection rates r (e.g. for the calculation of matching coefficients) may be chosen depending on the range of r-values occurring in the data set.
[a]Infection rates per unit per day.

those of fungal diseases when plotted over days for herbaceous annuals and years for woody perennials as appropriate time scales.

When Van der Plank (1963) introduced rates into epidemiology, he also analysed the effect host growth may have on rates as ρ (ro) for r (see also Sect. 4.1.3). Such an attempt was also made with data from field experiments for several pathosystems (Kranz 1975a,b) to account for the effect that host growth had on infection rates ρ. In the pathosystem *Rumex-Ramularia* for 16 out of 25 r-values, the corresponding values for ρ were higher by 0.02–0.15; nine were reduced. For *Hordeum-Erysiphe* five of eight ρ-values also increased with a maximum plus of 0.08. In contrast to these two pathosystems, host growth did not affect r-values of *Tussilago-Puccinia* and *Plantago-Oidium* significantly. Only two of six r-values had an increase in ρ. Hence, whether host growth affects the infection rates r either way (reducing or enhancing) seems to depend very much on the pathosystems. After a detailed discussion of some proposed equations to calculate the effects of host growth on rates, Waggoner (1986) found these effects less important. He quoted Kranz (1968b) and states: "...why labor to estimate precisely a shifty r that is a characteristic of the changing environment and not an immutable characteristic of a pathogen-host pair." For the seasonal course of epidemics, Waggoner (1986) considers the AUDPC as a better descriptor. However, its variability in relation to growth stages of hosts has to be taken into account, particularly when being used as a parameter for yield loss assessment.

In another field experiment over 3 years, infection rates of barley and wheat diseases were compared in a syn-epidemiological study (p. 16) after transformation with logistic and Gompertz functions (Khoury 1989). The rates, of course, varied between years and diseases (Table 5.8).

Rates (Table 5.8) obtained with the logistic function, with the exception of *Erysiphe* on spring barley (SB), were always higher than those calculated with the Gompertz function, but the significance of their R^2 was the same. Berger (pers. comm.) found this comparison between logistic and Gompertz peculiar, as the range of gompits is much less than for logits. Rates tend to be lower in spring

Table 5.8. Comparison of infection rates[a] of 98 disease progress curves of seven pathosystems[b] on spring (SB) or winter barley (WB) and spring (SW) or winter wheat (WW) in 3 years comparing logistic and Gompertz transformation. (Khoury 1989)

Pathosystem[b]		Year	N[c]	NO[d]	Logistic		Gompertz	
					Rate	R2 e	Rate	R^2
E. graminis	SB	85	1	10	0.03	0.67	0.003	0.65
		86	2	11	0.08	0.65	0.013	0.72
		87	5	11	0.07	0.80	0.010	0.89
D. teres	SB	86	2	11	0.01	0.31 ns	0.001	0.31 ns
		87	5	11	0.08	0.92	0.001	0.90
	WB	86	2	9	0.007	0.27 ns	0.001	0.27 ns
		87	5	8	0.15	0.65	0.006	0.62
P. hordei	SB	85	1	6	0.34	0.96	0.07	0.94
		86	2	7	0.24	0.90	0.04	0.84
		87	5	9	0.18	0.97	0.02	0.97
	WB	86	2	8	0.18	0.89	0.03	0.86
		87	5	5	0.09	0.99	0.02	0.95
R. secalis	WB	86	2	9	0.04	0.78	0.006	0.75
		87	5	8	0.15	0.85	0.02	0.84
E. graminis	SW	85	1	11	0.006	0.14 ns	0.001	0.10 ns
		87	5	13	0.11	0.83	0.006	0.81
	WW	86	2	12	0.31	0.77	0.004	0.75
		87	5	13	0.36	0.88	0.004	0.89
S. nodorum	SW	85	1	11	0.29	0.88	0.003	0.87
		87	5	13	0.11	0.93	0.02	0.91
	WW	86	2	12	0.06	0.79	0.008	0.76
		87	5	13	0.09	0.93	0.015	0.87
P. recondita	SW	85	1	10	0.09	0.62	0.014	0.61
		87	5	8	0.15	0.91	0.02	0.89
	WW	86	2	3	0.15	0.94 ns	0.02	0.98 ns
		87	5	9	0.31	0.96	0.06	0.88

[a] Infection rates are for weekly intervals; only maximum values are listed.
[b] The pathogens are *Erysiphe graminis* f. sp. *hordei* or f. sp. *tritici*, *Drechslera teres*, *Puccinia hordei*, *P. recondita* f. sp. *tritici*, *Rhynchosporium secalis*, and *Stagonospora nodorum*.
[c] Number of disease progress curves/year.
[d] Number of the weekly assessment dates.
[e] Most coefficients of determination R^2 for the rates are significant at $P>0.05$ unless marked ns.

barley and wheat than in their winter forms, except for barley leaf rust (*Puccinia hordei*).

When rates of epidemics are compared, the curves are usually linearised. However, the shape of original, non-linearised curves may contain biologically valuable information which makes them often more telling for the interpretation of results. Figure 5.4 compares curves for coffee berry disease (*Colletotrichum coffeanum*); a non-linearised disease progress curve (DPC) for untreated plants (upper curve) and plants treated with a fungicide (lower curve). Although the incipient r values (0.28) are the same, they do not differ over the entire epidemic with r=0.12 for untreated and 0.11 for treated plants respectively. The early sprays had a Δt-effect. The DPC of the treated one started later and crossed the 5% incidence level later. After this level, both curves start to soar with nearly the same r-values during the next interval and little differences thereafter. The control apparently becomes less effective during later stages of the epidemic (see also Table 5.4).

The phenomenon of higher r-values in the very early phase of an epidemic is of interest, particularly when it starts with low y_0 (Table 5.4, Fig. 5.4). Faster rates of epidemics with lower y_0 have been noted for several pathosystems. Such responses were also observed in comparative field experiments with *Cercospora arachidicola* on groundnuts, *Botrytis cinerea* on begonia, and *Uromyces appendiculatus* on Phaseolus beans with three levels of y_0 (Plaut and Berger 1981). In simulations with *Cercospora apii* on celery, assuming different initial disease intensities, r-values increased when y_0 decreased (Berger 1975, 1988), as in Table 5.4. Hence, more rapid rates compensated for the reduced y_0 and, consequently, y_{max} was not much different among the epidemics started with different y_0. Berger (1988) quotes Gregory et al. (1981) and Rouse et al. (1981), who independently had similar results with *Bipolaris maydis* and *Erysiphe graminis* f. sp. *tritici*. Rapid increase in incidence also compensated for low y_0 of *Septoria apiicola* on celery (Mudita and Kushalappa 1993). In model computations, it turned

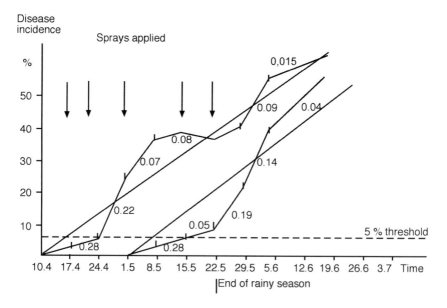

Fig. 5.4. Original and linearised disease progress curves of coffee berry disease (CBD) for untreated (*upper curve*) and treated with orthodifolatan (*lower curve*). The rates r between dates of disease recording show distinct difference along the two disease progress curves, while the r-values of the entire epidemic are practically the same, i.e. r=0.12 and 0.11 for untreated and treated, respectively. (Adapted from Steiner 1973)

out that underestimation of disease intensities near 0 when entered into equations underrated the actual r-values (Hau et al. 1989; Neher and Campbell 1992). For disease control such errors in disease assessment implicate wrong decisions, particularly when they are based on control thresholds. If it is not a statistical artefact, the more rapid increase in epidemics, i.e. higher rates, in their early stages with low y_0, as in Table 5.4 and Fig. 5.4, could be a principle. If so, there remains a risk: very low disease severities y_0 are easily overlooked. However, when these low severities are seen, they then tend to be overestimated with possible effects on y_{max}.

Among the factors affecting rates, i.e. climatic factors, host resistance, density of inoculum, there is also the mean age structure of lesions in a field (Table 4.12). The younger the mean age of lesions with their high sporulation intensity in a field, the higher the rate of disease progress. However, fast rates with early steep slopes of epidemics may lead more rapidly to a higher proportion of old lesions in pathogen populations. The inoculum consequently dwindles and the epidemic flattens or declines (Rangkuty 1984).

5.2.4 The Asymptote and Decline of Disease Progress Curves

After a progressive phase (d_p, Fig. 5.2) with a more or less steep increase in disease intensity, the progress curves may reach y_{max}, the maximum disease intensity, either in a peak, plateau or asymptote. If the increase in disease intensity slows down after an inflection point, the progress curve levels out to a plateau or an asymptote L (Vanderplank 1982). An asymptote, by definition, continues infinitely as an "equilibrium between host and pathogen" (Seem 1988). However, if this equilibrium is limited in duration, then the proper term for this part of the disease progress curve would be a plateau. The duration of plateaus varies among and within pathosystems. Examples for asymptotes are apple scab (*Venturia inaequalis*) on leaves (Analytis 1973) and *Oidium heveae* on leaves of the rubber tree (Populer 1972). These asymptotes occur because young leaves of their hosts are susceptible only for a short period after unfolding and then remain in the canopy until natural defoliation. A plateau ends with a decline in disease intensities at t_2 (Fig. 5.2) and a unilateral progress curve then becomes bilateral. The criteria applicable to asymptotes and decline are listed on page 98. Major factors that cause a progressive epidemic to turn into an asymptote (or plateau) are: (1) host growth and development with a change in phase-dependent resistance (Sects. 4.1.1 and 4.1.2) reflected in the carrying capacity K (Sect. 4.1.3); (2) the sporulation capacity resulting from the age structure of lesions in a crop (Sect. 4.3.2); and (3) the seasonal course of weather factors (Sect. 4.3.1).

It is common to set maximum disease intensity $y_{max}=1$ in equations for infection rates and equal to the carrying capacity K=1. This may be appropriate for disease incidence (DI) which often reaches y=100%, then being $y_{max}=1$. Disease severity (DS), however, rarely attains this intensity (Table 5.9). In a comparative field experiment with five pathosystems (Kranz 1975a,b, 1976, 1977), the maximum intensity was compared as disease severity per leaf (number of lesions per leaf, or percent leaf area covered by lesions), and incidence (number of leaves with at least one lesion). On the vast majority of leaves, the maximum severity DS_{max} (=y_{max}) was below 37% of the leaf area infected. Only leaves affected by powdery mildew (on *Hordeum* and *Plantago*) and the rust disease occasionally had a maximum severity in the range 37–100% (Table 5.9). The level of 37% disease intensity was chosen from personal experience and the Cobb scale. The implications for disease assessment are obvious: disease severities <37% deserve more attention in disease assessment and the design of classes and assessment keys.

Table 5.9. Maximum disease incidence (DI_{max}, % infected leaves) and maximum mean disease severity (DS_{max}, % per leaf), and percent leaves with <37% DS in five wild plant pathosystems. (After Kranz 1977)

Pathosystem[a]	Year	DI_{max}	DS_{max}	Percent leaves <37% DS[b]
Rumex–Ramularia	1969	85	7.9	96
	1970	93	11.6	93
	1971	100	5.7	100
Fragaria-Ramularia	1970	83	4.4	99
Hordeum-Erysiphe	1971	84	4.0	98
Plantago-Oidium	1970	91	15.6	70
	1971	74	52.8	42
Tussilago-Puccinia	1970	100	34.5	58
	1971	100	1.7	100

[a]*Rumex obtusifolius – Ramularia rubella, Fragaria vesca – Ramularia tulasnei, Hordeum vulgare – Erysiphe graminis f. sp. hordei, Plantago major – Oidium sp., Tussilago farfara – Puccinia poarum (0, I).*
[b]Rounded values.

For the calculation of rates realistic values of y_{max} or K should be used instead of 1 in the equation which assumes a possible 100% severity. Vanderplank (1982) suggested a conceptual approach to calculate the asymptote as $L \approx 1 - \exp(-\alpha L)$ where the progeny/parent ratio is $\alpha = iR$ per unit, i.e. the infectious period i with unlimited healthy tissue available. If a lesion produces one progeny only, i.e. $\alpha = 1$, like in endemicities (Sect. 5.1), Vanderplank in his Table 10.1 calculates the asymptote L to be 0 (meaning that the disease is near the limit of 0), $\alpha = 2$ equals L=0.797 or 79.7% disease intensity, with $\alpha = 4$, L would be at 0.98 and with $\alpha = 20$, about 0.99 (99%). Only with a high progeny/parent ratio α, y_{max} can be set as 1 in equations for the calculation of infection rates. Apparently, Van der Plank (1963) had tacitly assumed such high values.

Jeger (1987) computes the asymptote as $dK/dt = r_h(K-Y)(1-K/K_{max})$ when host and disease both grow logistically, with the carrying capacity K, defined as the maximum potential disease intensity to be carried, Y the intrinsic measure of disease, K_{max} the maximum attainable host growth, and r_h the rate of host growth. Under these conditions, if $r \geq r_h$ then y=Y/K will reach an asymptote of 1, but <1 when r<rh. The threshold theorem $iR \geq 1$, however, does not mean that disease cannot increase if iR<1 as disease severity does not depend on new lesions only, but also on the growth of the already existing ones (Hau 1988; Berger et al. 1997).

Vanderplank's equation $AS=1-(1-y_0)\exp(-iR\ x_{as})$ as formulated by Hau (1988) explains this better than the one proposed by Jeger (1986). Jeger added e^{y_0} to ensure that the asymptote equalled y_0 if $iR=1$. Later, Jeger and van den Bosch (1994a) agreed with Hau (1988) that, for asymptotic results of a finite plant population, an epidemic can proceed with $iR<1$. This depends on what values for i and R are chosen, close to <1 for a given y_0. However, the equations for asymptotes L, when there is infinite growth of susceptible host tissue (Vanderplank 1982), differ from plant populations with finite growth. In plant populations with no limiting host tissue, two distinct patterns of epidemic behaviour for polycyclic diseases are possible. This depends on whether $iR<1$ or $iR>1$. With finite growth the literal meaning of the threshold criterion iR is not yet clear for Jeger and van den Bosch (1994a), particularly when y_0 is quite high.

In experiments with the foliar pathosystems in Table 5.9 the leaf area, absolute disease severity in cm^2/plant and the disease severity in percent/plant were recorded. From this information actual leaf mass for K and y_{max} (for the latter see Table 5.9) could be studied. Figures. 5.5 and 5.6 demonstrate these relationships for two of the pathosystems in Table 5.9. The effect of host growth and the avail-

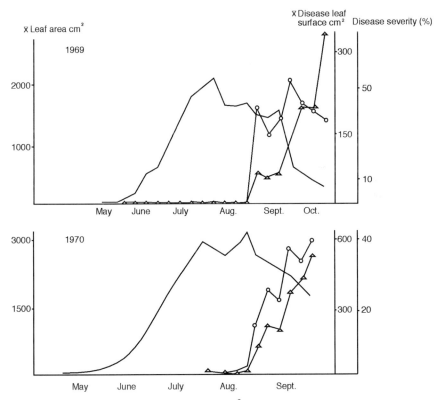

Fig. 5.5. Development of mean leaf area in cm^2 (*no symbol*), mean diseased leaf area in cm^2 (*circles*) and disease severity in percent (*triangles*) in the pathosystem *Tussilago farfara--Puccinia poarum* (0,I) in 2 years. (Kranz 1975a)

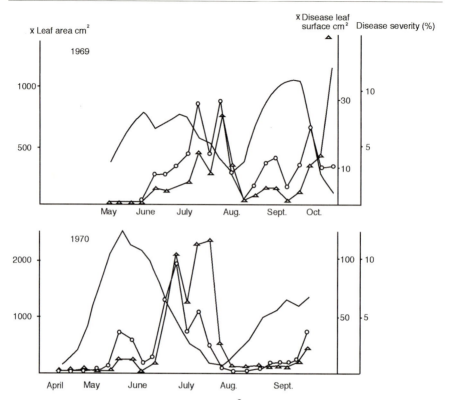

Fig. 5.6. Development of mean leaf area in cm^2 (*no symbol*), mean diseased leaf area in cm^2 (*circles*) and disease severity in percent (*triangles*) in the pathosystem *Rumex obtusifolius--Ramularia rubella* in 2 years. (Kranz 1975a)

able carrying capacity (sites for infection) in these experiments was discussed by Campbell (1998), who wrote "In most cases, the plot of rate of change diseased leaf area over time paralleled that of disease severity and the rate of change of disease severity equalled the rate of change of diseased leaf surface area minus the rate of change in leaf area or $r_{ds}=r_{dls}-r_{la}$. He (Kranz 1975a) demonstrated empirically that the rate of disease change corrected for host growth was equal to the sum of the rate of change of leaf area plus the rate of change of disease severity, or $r_{cg}=r_{la}+r_{ds}$, as expected for the proposed model of Vanderplank, and that r_{cg} can be negative even with a positive value of r_{ds}, if r_{la} is negative". The above equation applies when both host and disease increase logistically.

Neher and Campbell (1992) proved the advantage of a realistic carrying capacity $K_{max}<1$ for disease intensities actually recorded than setting $K_{max}=1$, as this has implications on the calculation of infection rates r. In a posterior model computation with data of 11 epidemics from various sources, they approximated their DPCs with the monomolecular, logistic and Gompertz functions. They estimated the effect of these functions on the infection rate which they designate as r* when values of $K_{max}<1$. As values of K_{max} decreased, r^* was increasingly underestimated particularly with lower K_{max} values, higher rates and longer duration of the epidemic. This effect was more pronounced when DPCs were approximated by the monomolecular function, but less so with the Gompertz and the logistic ones.

Neher and Campbell (1992) calculated disease progress curves using various parameter values for standardised mean rates r^*, y_0 and K_{max} for epidemics of a duration of 30 and 45 days. The monomolecular function was less sensitive than the Gompertz and logistic functions to changes in standard infection rates r^*. When epidemic duration was assumed to be 45 days instead of 30 days, effects of decreased K_{max} and increased standardised mean rates resulted in greater underestimation of r^*. This was less for the monomolecular and more for the Gompertz and logistic functions (Fig. 5.7). The underestimation of r_M (rate based on the monomolecular function) was affected more by decreasing K_{max} than by increasing r^*; the opposite was true for r_L with the logistic function. Changes in K_{max} and r^* affected the underestimation similarly in the Gompertz function. When the rates in the curves were calculated from empirical data with $K_{max}=1$ compared with actual K_{max} they decreased linearly with greater estimated K_{max} (Fig. 5.7).

These conclusions confirm empirical estimations by Analytis (1973) and Park and Lim (1985) that calculations with $K=1$ affect the value of r in the case of an actual asymptote below 100% disease intensity. Hence, $K=100\%$ disease intensity should not be used indiscriminately in equations to compute rates. Since K_{max} affects rates, Neher and Campbell (1992) suggested for $K_{max}<1$ shape parameters m with the values 0, 1 and 2 for the monomolecular, Gompertz and logistic functions respectively to use weighted mean absolute rates $\rho=r\ K/(2\ m+2)$ with $K=K_{max}$ and the rate of disease progress r calculated by the function used to fit the curve. This approach introduces the shape parameter m as a measurement term and descriptor (Table 5.13). Parameter K can be dropped from the equation, (if different with the same m), then rK could be calculated as an overall (mean) measure of the absolute rate (Neher and Campbell 1992).

It is evident from Fig. 5.7 that the parameter value of K_{max} affects the estimate of rates. However, more comparative research is necessary for a consolidated concept and theory. To resolve the problem of underestimation of the disease progress rate due to the use $y_{max}=1$ for maximum disease when actually $y_{max}<1$, Neher and Campbell (1992) derived some recommendations from their comparison by model computation (see also Campbell 1998): one may use the actual value of $y_{max}=K$ for linear regression when the actual y_{max} is known, or adopt more frequently occurring y_{max}–values of the disease in an area. For particular pathosystems, the most common y_{max} observed could be used, like the Cobb's scale for wheat stem rust in which 37% disease severity is rated as 100%=1. When estimates of K vary among individual DPCs a weighted absolute rate p (p. 124) should be used for comparisons among epidemics (Campbell and Madden 1990). Neher and Campbell (1992) feel that for the time being, too little information is available to classify pathosystems with regard to potential K_{max} when based on biological criteria only.

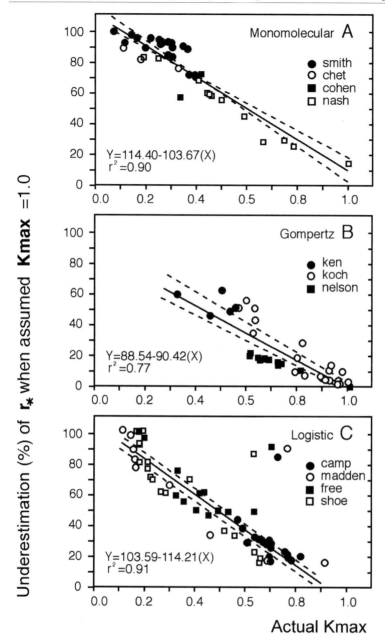

Fig. 5.7. Linear regression between percent underestimation of r*, the rate of disease increase if it is assumed that $K_{max}=1$, and the actual K_{max} values (with data from various authors as indicated by *symbols*, see Neher and Campbell 1992, Table 1). These regressions were fit to three growth functions shown in **A**, **B**, and **C**, with the 95% confidence intervals around the regression lines with indication of data sources. (Modified from Neher and Campbell 1992)

For details on the analysis and conclusions, the paper by Neher and Campbell may be consulted. Gilligan (1990b), in his model computation with some pathosystems, found marked differences in epidemics when control of disease was affected by reduction in the carrying capacity K or by delaying the onset t_0 of an epidemic. The asymptote or plateau obviously need more comparative across-studies similar to the two studies presented here.

Decline of disease progress curves after a peak or plateau is common in epidemics of pathosystems on wild host plants, perennials, and may also occur on crops before harvest. In general, the causes of declining disease intensities are either (1) the dilution of infected leaves by refoliation or (2) new growth of non-infected plant parts, an effect which can be enhanced by dwindling inoculum and increased resistance; and (3)defoliation by disease or aging leaves (Table 3.10), or a combination of all three factors (Kranz 1975b, 1976). Of course, host growth and defoliation are affected by environmental factors, e.g. adverse weather, the end of the growing season. For a more general discussion of effects of defoliation see Waggoner (1986 p. 108). In 3 years of field experiments with the five pathosystems in Table 5.9, there was no evidence for a decline due to the dilution of disease severity by new host growth. Under the conditions of the experiment with the pathosystem in Table 5.9, decrease in disease severity and decline of the disease progress curves studied were due rather to dwindling inoculum coupled with a thinning-out of diseased leaves. Defoliation of diseased leaves thus contributed to the decline of the disease progress curves (Table 5.10).

Table 5.10. Mean number of lesions/plant lost due to defoliation of diseased leaves early (E), halfway (M) and at the end (F) of the epidemics in 3 years (Kranz 1975b)

Pathosystem	Mean number of lesions lost in the years								
	1969			1970			1971		
	E	M	F	E	M	F	E	M	F
Fragaria-Ramularia	0.1	12	28	0.5	3	34	–	–	–
Rumex-Ramularia	–	–	–	0.5	6	14	0.4	0.3	11
Plantago-Oidium	3	46	167	0.4	0.5	0.5	0.07	0.04	0.05
Tussilago-Puccinia	5	1	12	0.1	111	219	4	11	11

Dwindling inoculum (Kranz 1976) also results from loss of diseased plant parts (Table 5.10) and increasingly less fertile lesions, i.e. too many old, "removed" lesions (Table 4.12). A relative reduction of inoculum may also be due to greater adult plant resistance. More inoculum is needed to produce a lesion on a more resistant tissue and this lowers disease efficiency. For example, *Erysiphe graminis* on spring barley on the first two leaf insertions had a disease efficiency of 20%, i.e. 5 conidia caused one lesion, but 50–60 conidia were needed to produce one lesion on the flag leaf (Aust 1981), thus enhancing the dwindling of available inoculum. Dwindling inoculum with less disease efficiency then causes the DPC to decline before harvest. The relative importance of either of the factors involved depended on the pathosystem and may differ with other conditions, e.g. weather (Kranz 1976). In any case, disease stops increasing "...not because there was no more healthy tissue available to infect, but because there was no inoculum left to infect it. The epidemic ran out of spores" (Vanderplank 1982).

Four pathosystems in Table 5.10 lost a higher proportion of lesions at the end of the epidemic or season by shedding leaves than during the epidemic. An exception is the pathosystem *Plantago-Oidium* in 2 out of the 3 years. The differences in numbers of lost lesions among years were rather well correlated with various disease severity. The relationship between leaves and lesions lost was significantly correlated in a Chi^2 test only for the bimodal pathosystem *Rumex - Ramularia* and in 1970 also for the unilateral *Fragaria - Ramularia*. However, even with declining leaf area, disease severity in terms of absolute leaf area diseased and disease severity in percent still could increase, e.g. *Fragaria-Ramularia* in 1970, and in *Plantago-Oidium* and *Tussilago-Puccinia* in all 3 years (Figs. 5.5 and 5.6). This paradox is explained by continuous growth of lesions on remaining leaves. These results suggest that the physiological age, and not the disease severity, had the overriding effect on defoliation. The duration of disease on leaves either did not differ between healthy and diseased leaves or disease could also enhance the life span of leaves (Table 5.11). Hence, disease endured on diseased leaves.

Table 5.11. Mean life span of healthy and diseased leaves (in weeks) and the mean number of lesions (N) or disease severity (percent) per leaf at the time each diseased leaf died or dropped. (Adapted from Kranz 1976)

Pathosystem	Year	Mean life span of leaves in weeks		Mean disease intensity	
		Healthy	Diseased	N	(%)
Rumex-Ramularia	1969	6±2	5±2	62.6	
	1970	5±2	4±2	14.7	
	1971	4±1	5±1	21.3	
Fragaria-Ramularia	1969	10±4	14±3	18.5	
	1970	9±3	11±2	33.0	
Plantago-Oidium	1971	5±1	6±2		22.5
Tussilago-Puccinia	1970	6±2	9±2		34.6
	1971	4±1	6±1		12.9

Some variables assumed to have an effect on defoliation were submitted to a multivariate analysis (Table 5.12), viz.: (1) duration of the epidemic (weeks); (2) duration of disease of each leaf (weeks); (3) number of lesions or disease severity (%) at the time of defoliation; and (4) disease severity 2 weeks prior to the date when leaves died or dropped. Only duration of the epidemic of both powdery mildews, or duration of disease on each leaf of the two *Ramularia* spp. explained the shedding (or decaying) of leaves significantly at $P>0.05$. For the pathosystem *Tussilago-Puccinia*, it could either be the duration of the epidemic, or the duration of disease on leaves as in 1970 when severity of disease was high (Table 5.10). In the multivariate analyses, the other variables, (3) and (4), always had low R^2 values, ranging from 0.0002 to 0.11 for (3), and from 0 to 0.05 for (4). As Waggoner (1986) argues, disease severity y actually decreases with time despite ongoing infection, if the rate of defoliation $1/t_d$ is faster than the rate r for increase in disease. Defoliation was not the only cause for the declining disease severity of alfalfa leaf spot (*Leptosphaerulina briosiana*) towards the end of an epidemic (Thal and Campbell 1988). Although bilateral DPCs are common for disease severity, defoliation in most cases is practically linear if the rate of defoliation $1/t_d$ is faster than the rate r for increase in disease. Of 59 pathosystems of wild host plants whose

Table 5.12. The variability in percent ($R^2 \times 100$) of two factors, viz. duration of the epidemic (in weeks) and duration of disease on a leaf (in weeks), that affected the shedding of diseased leaves in five pathosystems studied at the same site. (Kranz 1976)

Pathosystem	Year	Number[a] leaves	Duration[b] of the epidemic	Duration[b] of disease on leaf
Rumex-Ramularia	1969	102	7.57	82.90
	1970	223	0.51	12.76
	1971	298	0.20	33.29
Fragaria-Ramularia	1969	16	67.67	24.39
	1970	82	8.57	24.28
Hordeum-Erysiphe	1971	20	57.59	0.92
Plantago-Oidium[c]	1971	82	69.47	3.65
Tussilago-Puccinia[c]	1970	33	0.0	43.35
	1971	75	56.79	18.98

[a]Number of leaves included in the multivariate regression analysis.
[b]Mean duration of epidemics and disease on each diseased leaf was in weeks (mean values).
[c]Disease severity in percent, the others in mean numbers of lesions/leaf.

epidemics were studied on site for 2 years (Kranz 1968b), defoliation caused a complete destruction of host plants in about 16% of the pathosystems.

5.3 Unilateral Disease Progress Curves

Most disease progress curves are published as unilateral curves representing their temporal dynamics in somewhat varying shapes or patterns (Fig. 5.1). Behind these graphs of behaviour are interactions within the disease square and the internal "checks and balances". Unilateral DPCs often are incomplete DPCs because crops are harvested before a disease can complete its full course, i.e. there will be no degressive leg of the curve that may occur naturally. Complete bilateral DPCs are common in epidemics on wild host plants. Unilateral progress curves are often conceived as being S-shaped. Practically, however, S-shaped curves are more common for epidemics of some diseases, like late blight of potatoes, stem rust of wheat and less common for many other pathosystems, for instance apple scab, ear blotch of wheat, northern leaf blight of maize. The criteria for comparisons of unilateral curves are defined on p. 97.

Apparent deviations from the ideal S-shaped curve of *Phytophthora infestans* on potato (Fig. 5.8), commonly with S-shaped curves, can be attributed to specific changes in the time t_o, the onset of the epidemic, by weather and farming practices, or limited periods of susceptibility of the host plant including control measures.

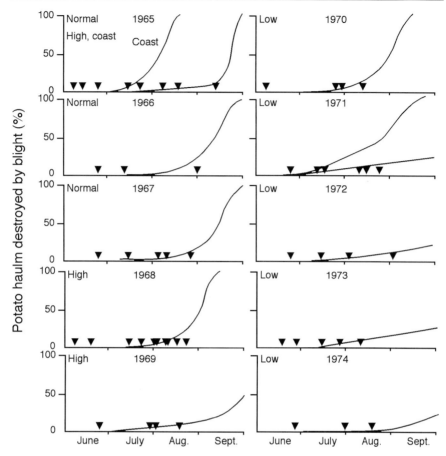

Fig. 5.8. Actual disease progress curves of late blight (*Phytophthora infestans*) in 10 years in the UK at various disease intensities. (Modified from Croxall and Smith 1976)

In addition, regional variation can be observed in disease intensity. Late blight was less severe in wetter areas in the west and cooler areas in northern England than in the east, where the national mean intensity was exceeded every year (Croxall and Smith 1976). Epidemics of *Mycosphaerella graminicola* (*Septoria tritici*) on wheat normally have unilateral DPCs. However, in one season with continuous cool weather, the progress curve decreased and turned bilateral (Hart et al. 1984). Similar disease progress to fungal pathosystems is reported for virus pathosystems, e.g. sugar beet yellow virus (Fig. 5.9), mosaic on cantaloupe and others (Fig. 5.10). There is some good correlation between disease progress (taking into account a shift due to latency) and vector density in the virus diseases presented, which Thresh (1983) saw as a more general phenomenon.

Van der Plank (1963) first defined and compared simple and compound interest diseases whose disease progress curves could be linearised by the monomolecular and logistic growth functions. In addition, Analytis (1973) applied the Gompertz,

Mitcherlich and Bertallanffy-Richards growth functions to the epidemics of apple scab. Later, the versatile Weibull growth function (Campbell et al. 1980) and an exponential and log-log functions (Campbell and Madden 1990) were introduced.

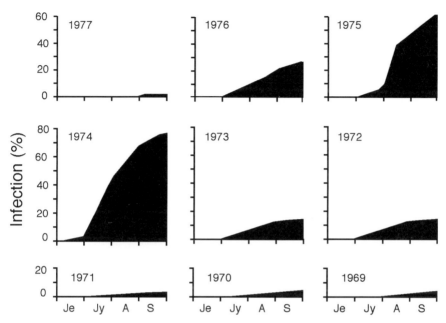

Fig. 5.9. Mean monthly incidence of sugar beet yellows in England from 1969 to 1977. Data for representative crops throughout the beet-growing area sown annually in March or April. (After Thresh 1983, with information from G.D. Heathcote)

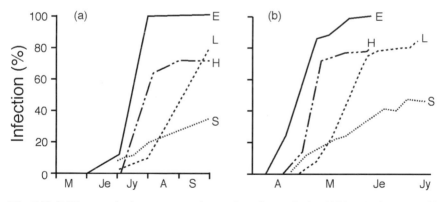

Fig. 5.10. Differences in the amount and sequence of spread of aphid-borne viruses at different sites, seasons and epidemic courses. **a** The spread of beet virus yellows at different sites in England. **b** The spread of mosaic viruses of cantaloupe in Arizona in different years. *E* Spread early in the growing season, *L* late onset and spread, *H* early onset, but halted spread late in the season, *S* slow spread throughout the season. (After Thresh 1983)

Table 5.13. Differential equation (dy/dt), integrated equation y, and linearised forms of growth functions in use for unilateral disease progress curves. (from Campbell and Madden 1990, with their designations)

Function		Equation
Exponential	dy/dt	$r_E y$
	y	$y_0 \exp(r_E t)$
	Linear	$\ln(y) = \ln(y_0) + r_E t$
Monomolecular	dy/dt	$r_M(1-y)$
	y	$1 - [(1-y_0) \exp(-r_M t)]$
	Linear	$\ln[1/(1-y)] = \ln[1/(1-y_0)] + r_M t$
Logistic	dy/dt	$r_L y(1-y)$
	y	$1/[1 + \exp(-(\{\ln[y_0/(1-y_0)] + r_L t)\})]$
	Linear	$\ln[y/(1-y)] = \ln[y_0/(1-y_0)] + r_L t$
Gompertz	dy/dt	$r_G y[-\ln(y)]$
	y	$\exp[\ln(y_0)\exp(-r_G t)]$
	Linear	$-\ln[-\ln(y)] = -\ln[-\ln(y_0)] + r_G t$
Log-logistic	dy/dt	$r_{LL} y(1-y)/t$
	y	$1/[1 + (1-y_1)/y_1]t^{-r_{LL}}$
	Linear	$\ln[y/(1-y)] = \ln[y_1/(1-y_1)] + r_{LL}\ln(t)$
Bertalanffy-Richards	dy/dt	$r_R y(1-y^{m-1})/(m-1)$
	y	$[1-B\exp(-r_R t)]^{1/(1-m)}$ if m<1 (1)
		$B = 1 - y_0^{1-m}$
		$[1+B\exp(-r_R t)]^{1/(1-m)}$ if m>1 (2)
		$B = y_0^{1-m} - 1$
	Linear	$\ln[1/(y^{1-m}-1)] = -\ln[1/(y_0^{1-m}-1)] + r_R t$ (1)
		$\ln[1/(y^{1-m})-1] = \ln[1/(y_0^{1-m})-1] + r_R t$ (2)
Weibull	dy/dt	$(c/b)[(t-a)/b]^{c-1}\exp-[(t-a)/b]^c\}$
	y	$1-\exp\{-[(t-a)/b]^c\}$
	Linear	$\{\ln[1/(1-y)]\}^{1/c} = -a/b + t/b$
		or $\ln\{\ln[1/(1-y)]\} = -c\ln(b) + c\ln(t-a)$

Since then, growth functions have been applied for the mathematical description and approximation of observed disease progress curves in epidemiology (Table 5.13): the exponential (logarithmic sensu Van der Plank), monomolecular, logistic, Gompertz functions (Fig. 5.11) have three or fewer parameters (Campbell and Madden 1990). Their parameters as criteria for comparison are initial disease intensity y_0, disease intensity y_t at any date of recording, a rate of change r, and the maximum disease intensity y_{max} or K, the carrying capacity and time t which is implicit in growth functions.

Functions with a shape parameter m (Table 5.13) are the Bertalanffy-Richards and Weibull functions, to some extent also the log-logistic (where shape cannot be separated from the rate). Shape parameters, though their choice may be arbitrary, facilitate the approximation of disease progress curves. In addition to m, as in the Bertalanffy-Richards function, the Weibull has a scale parameter b, and a shape parameter c. The parameter a (t_0 in Fig. 5.2) stands for the start of the epidemic in time units, b is a scale parameter (in time units) and c describes the kurtosis of the curve. Extensive treatments of the mathematics of growth functions applicable to DPCs are by Madden (1986), Waggoner (1986), Gilligan (1990b), Madden and Campbell (1990) and Campbell and Madden (1990).

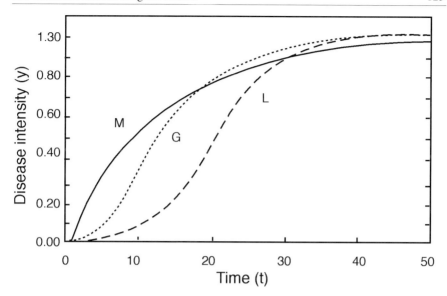

Fig. 5.11. The theoretical course of epidemics after approximation with the growth functions monomolecular (*M*), logistic (*L*) and Gompertz (*G*), see also Fig. 5.3. (Adapted from Campbell 1998)

Table 5.13 lists growth functions used in epidemiology. These functions are based on the following assumptions (Campbell and Madden 1990): (1) the environment in which the epidemic develops is uniform; (2) the population of hosts and pathogens is genetically uniform with respect to susceptibility and virulence/aggressiveness, and age classes; (3) there is a constant host area to be infected; (4) the change in disease intensity is continuous and immediately visible; (5) the spatial pattern of disease within and among hosts is random.

For the exponential growth function (Fig. 5.3; Table 5.13) dy/dt is proportional to the disease as long as sufficient disease-free host is available and no constraints impede the progress. This function in polycyclic diseases, according to Van der Plank (1963), strictly applies to the early stage of an epidemic only, up to 5% disease intensity or y=0.05. The monomolecular function describes growth of monocyclic diseases (simple interest diseases) in relation to the disease-free host with dy/dt decreasing negatively and exponentially with time. It is often useful for DPCs of monocyclic diseases. A logistic function is appropriate for polycyclic diseases with increasing multiple infections. Because of decreasing disease-free host a correction term (1–y) is included. The log-logistic function is a generalisation of the logistic function directly proportional to y and 1–y. It is inversely proportional to t over time. With the Gompertz function, one more aptly describes epidemics with more rapid development in their early stages than with the logistic function.

Skewness m of the disease progress curve (Table 5.3) in the growth functions of Bertalanffy-Richards and Mitscherlich (Analytis 1973) may be a good criterion for comparison. That is, it may describe the natural coincidence between susceptible stages of the host, the amount of inoculum present in the course of an epidemic or both. Analytis (1973) found the shape parameter in the Bertalanffy function a rather stable criterion if t_0 (Fig. 5.2) is related to t_B, the beginning of host vegetation, flowering etc., and temperature. The Richards function is a generalised Bertalanffy function. The shape parameter m makes it a rather versatile function. The shape parameter m=0 stands for a monomolecular, m=1 for a Gompertz and m=2 for a logistic epidemic. When fitting curves of the Gompertz type with a Richards function near m=1, the problem may arise of how to choose between the two functions in Table 5.13 (Berger, pers. comm.). The Weibull function is even more universally applicable to fit DPCs with the inflection point for c>1 at $y=1-\exp(1/c)^{-1}$ and $t=b[(c-1)/c]^{1/c}+a$. The shape parameters m in the Bertalanffy-Richards and c in the Weibull functions are useful to analyse disease progress although the parameters m and c cannot be explained specifically like the biological implications of y_0 and r. Nevertheless, these parameters may indicate some biological facets (e.g. juvenile or adult plant resistance) or the effect of some environmental factors. For this reason, the original plotted curve should be examined when interpreting the curves obtained with growth functions with m and c. The biological meaning of these curve parameters may well be a rewarding research proposition. Bergamin Filho et al. (1998) propose a "more-than exponential" function $dy/dt=(r+bt)y$ and its integrated form, $y=y_0\exp(rt+bt^2/2)$. With this function, they amended the exponential function in Table 5.13 with b as a parameter, to approximate the progress curves of the apparently abiotic fatal yellowing of oil palms. No other biological process is known so far to follow this function.

Appropriateness of a function to describe disease progress data can be tested by the coefficient of determination (R^2), residuals or mean square error (MSE). To compare values of R^2 and MSE precisely among functions with different dependent variables, for example $\ln[y/(1-y)]$ and $\ln[1/(1-y)]$, R^2 should be recalculated after back-transformation with predicted disease values on the observed disease values. If two curves are well described by the same function, values of r and y_0 can be compared statistically by a t-test or a 95% confidence interval about the difference between the parameter values (Reynolds and Neher 1997; Campbell 1998). For multiple disease progress curves from a designed experiment with replicated treatments, the estimates of r and y_0 of the curves can be compared with replicated treatment differences of ANOVA (Campbell 1998). However, if the curves are best described by different functions, a weighted mean absolute rate of disease increase ρ (ro) (Campbell and Madden 1990) can be calculated as $\rho=r^*K/(2 m+2)$, where r^* is the rate parameter of the specific function, e.g. the monomolecular, K is the maximum disease intensity ($0<K\leq1$) and m is the shape parameter for the Bertalanffy-Richards function (see above).

For comparative epidemiology, the biological significance of the various parameters in growth functions is rather clear with the monomolecular, logistic and Gompertz functions. This is because disease intensity and time are the parameters. Gilligan (1990b) found the monomolecular function in a model computation generally less flexible than the logistic. If the Gompertz function provides a good fit to the DPC of epidemics with an early and rapid progress, the rate should be calculated with gompits $-\ln(-\ln y)$ instead of $\ln[y/(1-y)]$. For the Gompertz function, Waggoner (1986) sees a somewhat turbid rationale, with the Richards function various rationales and with the Weibull no clear biological rationale. The flexibility of the Weibull and the Bertalanffy-Richards functions warrants some caution when comparing epidemics. However, when parameter values are carefully selected, these functions will at least serve for curve fitting. The polynomial functions have no biological significance, but may be useful for the approximation of curves in models. Schrödter (1965) successfully used polynoms to establish a function of temperature equivalents.

The sensitivity of the three parameters of the Weibull function to changing factors was studied in simulation experiments by Thal et al. (1984). Data were generated for the study with monomolecular, Bertalanffy-Richards (for m=0.5), Gompertz and logistic growth functions. For this comparison, the following conditions were used: six levels of y_0 and ten levels of ρ (ro – a common weighted mean rate parameter), four recordings of disease intensity and four final disease levels (p. 98). Estimates of the parameters were sensitive to the weighted mean rate ρ when data were generated with the Gompertz and the logistic functions and when values of y_0 were high. Estimates of b of the Weibull function were inversely related to ρ and estimates of c were insensitive to changes in the rate. Estimated parameter values for all functions, except the monomolecular one, reacted to changes in y_0. A reduction of the final disease intensity y affected c, the magnitude of which increased as the inflection point of the generated DPC increased. Generally, estimates of the parameters a, b and c were highly correlated. This may indicate overparametrisation of the Weibull function. Consequently, when using functions with shape parameters, one cannot simply compare rates r among epidemics without regarding the shape of the curves (Campbell and Madden 1990).

Some experimental comparisons of growth functions obtained in pathosystems are referred to next. Epidemics of apple scab on leaves are S-shaped disease progress curves for about 6–7 weeks and an extended asymptote thereafter (Analytis 1973). These curves are best fitted by the Bertalanffy (Bertalanffy-Richards in Table 5.13) and Mitscherlich functions with the shape parameter m=2 (n=2 in Analytis 1973). After a short period of disease increase during which, if the severity of the scab is assessed at very short intervals, the increase in disease would follow an S-shaped curve. Thereafter, this curve becomes rather a monomolecular curve with a long asymptote (Fig. 5.3).

Monomolecular and Weibull functions fitted the simple interest phase of the naturally polycyclic pathosystem *Erwinia-Lycopersicon* (Berger and Bartz 1982). The DPCs of thrips-vectored tomato spotted wilt virus (TSWV) in groundnuts were well described by the monomolecular function in each cultivar and both years of field experiments (Camann et al. 1995). Essential was that disease incidence resulted from primary transmission and there was a limited secondary spread of TSWV after it had become established in the field. A comparison of seven genotypes of groundnuts under four different inoculum densities of *Cylindrocladium crotolariae* was conducted during 2 years of field experiments (Culbreath et al. 1991). Among the monomolecular, logistic and Gompertz functions, the logistic showed the highest R^2. After logit transformation with $y_{max}=1$ the rates of the DPC for incidence were similar in the moderately resistant genotypes, faster in the susceptible ones and very slow in the two resistant curves. (Culbreath et al. 1991). Most of the DPCs of common root rots of spring wheat and barley primarily caused by *Cochliobolus sativus* (*Bipolaris sorokiniana* = *Helminthosporium sativum*) fit the logistic function, or the versatile Weibull function (Stack 1980). The implication of this analysis is that secondary infection is involved in the rate of progress of this disease throughout the season with no detectable effects of ontogenetic changes in susceptibility and environmental factors. In field experiments between 1974 and 1980, Campbell et al. (1984) tested the monomolecular, logistic, Gompertz, Bertalanffy-Richards (n=2) and Weibull functions for their goodness-of-fit to the DPCs of 50 epidemics of tobacco black shank (*Phytophthora parasitica* f. sp. *nicotianae*). Based on R^2 and residuals, the logistic function was 28 times appropriate, the Gompertz 14 times. All of these and the following examples prove that epidemics of pathogens can be described by one or a few characteristic growth functions. Major determinants are, apart from weather, the development of the host and its resistance, and the epidemic competence, in particular its rate component.

In a posterior model computation, Gilligan (1990b) tested the effects of genetic, chemical and cultural methods on the control of seven pathosystems with four common parameters: viz. lower asymptote, upper asymptote, rate and delay, the vertical displacement or differing upper asymptote, or horizontal displacement, and rates. Two growth functions were eventually used. The logistic function provided the best fit to the polycyclic epidemics of *Phytophthora infestans* on potato, *P. cryptogea* on raspberry, *Fusarium oxysporum* f. sp. *lycopersici* on tomato and *Puccinia recondita* on wheat and triticale. The monomolecular function was applied to *Sclerotium rolfsii* on carrots and *S. cepivorum* on garlic. Gilligan (1990b) then performed a series of single parameter comparisons with implicit tests for a common curve, and finally examined hierarchal models with progressively fewer common parameters. He first fit the selected function separately to the disease progress curve as measured for each of the treatments and subsequently constrained one or more parameters. This was to make them fit all treatments, while the remaining parameters were separately fitted. The changes thus obtained were tested by the residual mean squares, and differences amongst treatments could be shown (Sect. 7.2). According to Campbell (1998), this study by Gilligan (1990b) proposed a general method of comparing non-linear DPCs as an interesting approach to the comparison of factors and parameters: "He... used the method of parallel curve analysis to examine whether or not one curve was sufficient to describe curves for all treatments and, if not, to test whether certain parameters were common among the treatments while others varied" (Campbell 1998).

Nelson and Campbell (1993b) studied diseases on white clover (*Trifolium repens*) and on tall fescue pasture (*Festuca arundinacea*) concurrently at the same site over 2 years at 16 dates as a comparative field experiment (see also p. 139). Disease incidence was studied for the following foliar diseases caused by: *Pseudomonas andropogonis*, *Cercospora zebrina*, *Curvularia trifolii*, *Rhizoctonia solani*, and *Stagonospora meliloti*, plus a complex of the viruses alfalfa mosaic virus (AMV), clover yellow vein virus (CYVV), and red clover mosaic virus (CCMV). The disease progress was non-monotonic and gave poor fits to the three growth functions, viz. monomolecular, logistic and Gompertz functions. These poor fits were probably due to new leaf growth and defoliation. No difference was detected between the course of the curves on virus-free or virus-infected plants. Both cultivars used in the experiments were similar with regard to the shape of the DPCs and y_{max} as well as of AUDPC.

Factors which cause a non-linear increase of disease are, for instance, ontogenetic changes in host susceptibility during the vegetation period, age distribution of lesions, the intermittent availability of inoculum, effectiveness of control or cultural measures and recurrent weather events.

Comparisons of weather factors and their impact on DPCs can best be measured by specially defined weather factors derived as "secondary factors" from the one measured with the instruments (Table 5.14; Sect. 4.3.1). Secondary weather variables can be more appropriately linked with relevant time intervals during epidemics and have been of advantage in modelling and for the development of forecasts.

Table 5.14. Secondary weather variables for cereal diseases[a]

Code	Definition and parameter of variables
MMAX	Mean maximum temperature (°C)
MMIN	Mean minimum temperature (°C)
MAVE	Mean daily temperature (°C)
XTT	Mean daytime temperature (°C)
XNT	Mean nighttime temperature (°C)
XT90	Mean temperature (°C) at relative humidity >90%
XTBB	Mean temperature (°C) during leaf wetness
PDD	Positive degree-days
DLOC	Sum days with mean temperature <0 °C
DG25C	Sum days with mean temperature >25 °C
ZF	Sum hours with relative humidity >90%
DBN	Sum hours with leaf wetness
TPREC	Sum precipitation in cm
CDWP	Sum successive days with rain
CDWOP	Sum successive days without rain
XRAD	Mean intensity of radiation (cal/cm^2/min)
XW	Mean wind speed (m/s)

[a]Compiled and selected from Coakley et al. (1988) and Khoury (1989). Coakley et al. refer their parameters to pentads (5 days), Khoury to weekly intervals. See p. 65 for primary weather factors.

The effects of weather variables defined in Table 5.14 were compared in a syn-epidemiological study on wheat and barley for weekly intervals of disease progress of *Erysiphe graminis* f. sp. *tritici* and f. sp. *hordei, Stagonospora nodorum,* and *Puccinia recondita* f. sp. *tritici,* as well as *Drechslera teres, P. hordei* and *Rhynchosporium secalis* (Khoury and Kranz 1994). The effect of these weather variables on both rusts of the two hosts are presented in Table 5.15. While the correlation coefficients of *P. hordei* mostly are significant, those of *P. recondita.* are not. This was because of the greater variability, although variables differ in the correlation coefficients with both *Puccinia* species.

Table 5.15. Correlation coefficients r for the relationship between defined secondary weather variables and the weekly infection rates for new infections of the pathosystems *Puccinia hordei* – spring barley and *P. recondita* – spring wheat in 1987. (Khoury 1989)

Weather variable	P.hordei	P.recondita
	r	r
Mean day temp. (°C)	0.70	0.69 ns
Mean night temp. (°C)	0.58	0.66 ns
Mean temp. during >90% relative humidity	0.45	0.56 ns
Mean temp. during leaf wetness	0.41	0.52 ns
Mean temp. without leaf wetness	0.74	0.60 ns
Temperature sum (per h)	0.65	0.64 ns
Number hours with temp. >7 °C	0.43	0.50 ns
Number hours with temp. >12 °C	0.57	0.57 ns
Number hours relative humidity <70%	0.73 ns	0.50
Leaf wetness duration (h)	0.61 ns	0.43 ns
Rain (mm/week)	0.70 ns	–
Mean intensity of radiation (cal/cm^2/min)	0.77	0.56 ns

The r-values without ns are significant at *P*>0.05.

To linearise unilateral curves a transformation procedure is a prerequisite to compute linear regressions and rates. The more appropriate transformation functions may have to be tested before they are applied (Hau and Kranz 1977; Analytis 1979). Thirteen disease progress curves of the five pathosystems in Table 5.9 studied for 3 years were linearised by means of eight transformation functions (Hau and Kranz 1977). None of these functions fit every one of the 13 curves equally well. It was the shape of the DPC only that determined the function for the most accurate transformation. It was also obvious that three or four of the eight functions tested for linearisation sufficed for the majority of disease progress curves studied. In the study these transformation were $\ln[1/(1-y)]$, $\ln[y/(1-y)]$, $\ln[1/(1-\sqrt{y})]$ and $\arcsin\sqrt{y}$.

5.4 Bilateral and Multimodal Disease Progress Curves

Bilateral epidemics, i.e. naturally completed during a season, have disease progress curves with both progressive and degressive legs. Their degressive phase $d_D=t_F-t_2$ (Fig. 5.2) can vary and disease intensity y_F will rather be closer to 0 than to y_{max}. Epidemics can also be multimodal and oscillating (periodic; Fig. 5.1d,e).These multimodal epidemics occur because of disease cycles, discontinuities in disease progress, variation in incubation periods, changes in host sus-

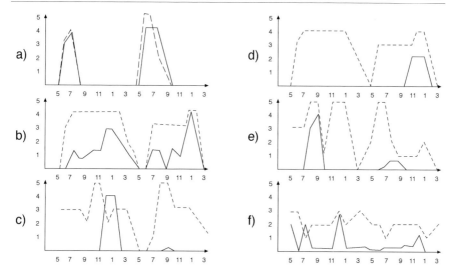

Fig. 5.12a–f. Epidemics (*solid lines*) on wild host plants in relation to foliage of host (*broken lines*) during months of two successive years in five nominal scales of intensity: **a** *Puccinia conclusa* on *Cyperus rotundus*, **b** *Puccinia versicolor* on *Andropogon tectorum*, **c** *Puccinia levis* on *Digitaria longiflorae*, **d** *Phyllachora fallax* on *Andropogon tectorum*, **e** *Phyllachora digitariae* on *Digitaria longiflorae*, **f** *Peronospora favargeri* on *Euphorbia hirta*. (Kranz 1968a)

ceptibility due to new growth or other factors. These types of progress curves are common in perennial crops and among pathosystems of wild host plants. Of 59 pathosystems investigated concurrently for 2 years in their natural habitats, practically all had complete, but diverse curves (Kranz 1968a). In Fig. 5.12, disease progress curves of six of these pathosystems are shown in relation to the development of the foliage of their host plants.

From Fig. 5.12, one can conclude that diseases rather tend to have discontinuous epidemics even in pathosystems with practically continuous host growth in terms of foliage (e.g. Fig. 5.12d,f). A continuous epidemic was recorded for only a few of the 59 pathosystems, like the practically perennial *Euphorbia hirta* (Fig. 5.12f). The disease progress, however, is largely independent of the amount of foliage, with one exception during the dry months from November through January of both years. However, a certain correlation with foliage available for infection was obvious in the pathosystems a, b, and f, whereas disease progress in c, d, and e does not show any relationship with host growth. For only two of the pathosystems in Fig. 5.12a,b were disease progress and available foliage correlated. In pathosystem a, the growth of host and disease intensity seemed to be concurrent. In pathosystem b, the epidemics reach their peaks towards the end of the host vegetation. In the four other examples in Fig. 5.12, the diseases followed their own rhythms. Hence, the amount of susceptible host tissue may or may not have a bearing on the development of epidemics, as was already pointed out in Section 5.2.4.

Within pathosystems patterns of disease progress curves may change because of the effect of factors under which they developed. In experiments for 3 years at the same site among the pathosystem in Table 5.12 the curves of *Plantago-Oidium* in 2 of 3 years were bilateral, and unilateral in one. *Tussilago-Puccinia* (Fig. 5.5) was linear in one and bilateral in the second year. The pathosystem *Rumex-Ramularia* (Fig. 5.6) had bimodal DPCs in all 3 years of the experiment. Progress curves of coffee leaf rust (*Hemileia vastatrix*) in Kenya may either be bilateral or bimodal, depending on the pattern of annual rainfall in a given year. Bimodal curves are recorded, even with little rain, up to 1800 m altitude. Above this altitude they invariably become bilateral (Bock 1962). Disease progress curves from *Peronospora farinosa* in both spinach and *Chenopodium album* are bilateral (Frinking and Linders 1986). On *C. album*, *P. farinosa* first caused slight peaks before the bimodal main epidemics start. In contrast, in spontaneous populations of *C. album*, logistic curves were also recorded. The infections in spinach come from sporangia, while the one on *C. album* arose from oospores. Of the two bilateral DPCs on *C. album* in two populations, one was negatively skewed, the other positively. The slower disease progress in the early phase of the epidemic was prompted by host growth, the decline of the curve at the end occurred when the total leaf area was decreasing and plants started to set seed. Also, a decrease in susceptibility was given as cause. The decline of the epidemic on *C. album* began before stem elongation and before flowering (Frinking and Linders 1986). Despite such variation in DPCs, there are pathosystems which can have a tendency for bilateral, bimodal or periodic progress curves, conditions permitting, while others are only known with unilateral curves. Hence, bimodal and oscillating (periodic) progress curves occur more often in certain diseases when conditioned by, for instance, a regular weather pattern affecting either host growth or development of the disease or growth flushes like in the pathosystem *Rumex-Ramularia* (Fig. 5.6). Bimodal epidemics occur in annual crops, but are more frequent in epidemics of perennial crops; The latter often being polyetic. Whether a pathosystem has a bilateral, bimodal or periodic pattern (Fig. 5.14) of disease progress can be decided only after more than 2 years of data recording.

A somewhat different type of multimodal epidemic may occur in crops when pathogens in sequence cause diseases on various plant organs. For instance, *Pyricularia grisea* incites seedling blast, leaf blast, neck blast and finally panicle blast with its own disease progress curves. *Phytophthora* spp. on rubber attack fruits first, then leaves and finally tapping panels on trunks.

Established functions for bi- and multilateral progress curves like the unilateral ones are rare. Appropriate methods to analyse and compare bilateral and multimodal epidemics may either be formal (most frequently) or biologically meaningful. Among the formal ones, "any continuous or reasonably discontinuous periodic function can be represented by a trigonometric polynomial with any degree of accuracy" (Batschelet 1971). However, their biological significance can be limited. It is an "...empirical fact, that it is but rarely that we find in practice an empirical distribution, which could not be satisfactorily fitted by any such curves" (Neyman 1939). Periodic data of oscillating, wave-like curves can be fitted by another trigonometric polynomial known as Fourier analysis, or with the ARIMA approximation (Pak 1992; Yang and Zeng 1992).

The area under a disease progress curve (AUDPC) is biologically more meaningful than polynomials and ARIMA. The AUDPC fits bilateral curves, cyclic
events when y is plotted against t, to discontinuities in disease progress. When di-
viding the total time duration for epidemics of AUDPC into different time intervals, possibly connected with host stages, a standardised SAUDPC may be used
(Navas-Cortés et al. 1998) either with units of percent-day or proportion-day. In
addition to AUDPC, Campbell (1998) proposed the analysis of variance with time
as a factor as a valid statistical description of disease progress (see Reynolds and
Neher 1997). Repeated-measure analysis of variance is another alternative when it
is not possible to analyse disease progress with growth functions.

Functions to describe bilateral curves can be determined iteratively with appropriate software on personal computers. Computer software now offers the possibility to identify functions to fit bilateral curves from experimental results, e.g.
exponential or power functions. Larios and Moreno (1977) obtained such functions when approximating the bilateral curves of *Oidium manihotis* and *Sphaceloma* sp. on cassava (Fig. 5.13). These pathosystems were studied side by side to
compare the effect of different cropping systems. Their curves for incidence and
severity follow a similar course. It would be worthwhile to know whether there are
functions of this type that differ among certain groups of pathosystems, or, at
least, have different types or classes of exponents.

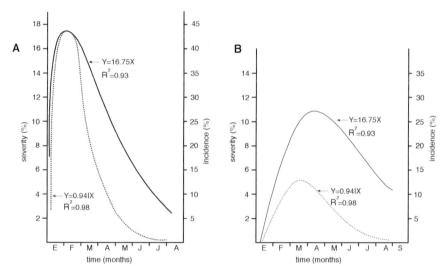

Fig. 5.13A, B. The course of epidemics with curves for severity and incidence of *Oidium
manihotis* and *Sphaceloma* sp. on cassava (*Manihot esculenta*) during a growing season
(January to September) with the functions fitting the four disease progress curves and their
multiple coefficients of determination R^2 (Larios and Moreno 1977). The *solid lines* indicates incidence, the *dotted lines* severity, both in percent

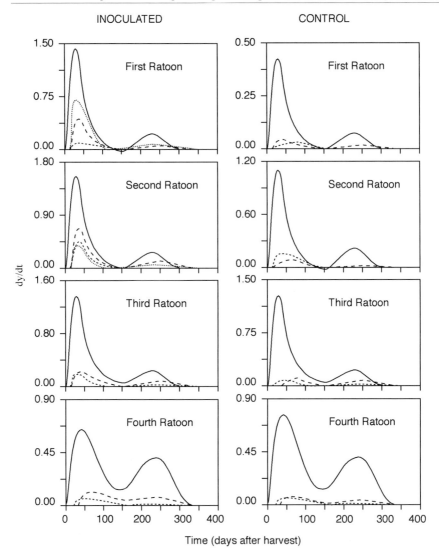

Fig. 5.14. Derivates of the generalised Gompertz function fitted to double sigmoid DPCs of sugarcane smut in inoculated and control plots for four successive ratoons and four cultivars. (Amorim et al. 1993)

In sugarcane smut (*Ustilago scitaminea*), each year has apical and lateral infections that lead to two waves of disease incidence of the cane (Amorim and Bergamin Filho 1991a,b). Hau et al. (1993) proposed double sigmoid growth functions either as two linked logistic or Gompertz functions or a combination of both to describe these curves. They tested ten common growth functions on data from epidemics of sugarcane smut and found that the generalised monomolecular and the generalised Gompertz functions, each with five parameters, were the most useful functions to fit the double sigmoid pattern of *U. scitaminea* with values for R^2 always above 0.95. Amorim et al. (1993) fitted the double sigmoid functions to annual progress curves, which differ in steepness in the early epidemic. They used the Gompertz function as parameters (and criteria) and also included the upper asymptote, mean infection rate, the transition time between growth waves, and the latent period to compare cultivars (Fig. 5.14). Both time-related curve elements, i.e. transition time and latent periods, did not differ among the curves. It was the upper asymptote, however, that differed significantly. However, when smut was plotted with the last cumulative intensity per year over 5 years, the polyetic DPCs in four cultivars were a typically unilateral DPC.

The temporal patterns of three whitefly (*Bemisia tabaci*) transmitted gemini viruses, viz. African cassava mosaic, okra leaf curl and tobacco leaf curl in West and East Africa and India, follow the annual temperature curve in correspondence with changing vector densities. All three diseases share some general characteristics as their epidemics are driven by the same key variables. Their spatial spread is characterised by strong border effects because whiteflies accumulate on the wind-exposed borders of the field (Fargette et al. 1993).

5.5 Disease Progress Curves of Soilborne Diseases

Methods of measurement and modelling of temporal disease progress of root diseases sometimes require approaches rather different from aboveground pathosystems and their epidemics (Gilligan 1994). Criteria particularly needed for their comparison, apart from those given on p. 97, time and distance, include : density of inoculum in appropriate terms, a parameter on survival and the amount of effective inoculum Q available at the (possible) start of an epidemic. However, also with soilborne pathogens, all components of the infection chain (Sect. 4.3) apply when specifically adjusted to the conditions of soil. Once soilborne pathogens are manifested on plant shoots, the same criteria and methods for comparisons as for foliar diseases can be used. Campbell (1986) treats disease progress curves of root diseases in some detail, including monitoring, analysis, factors affecting and interpretation of curves obtained. He also lists examples of the growth functions that have fit progress curves of soilborne diseases. These functions include the Bertalanffy-Richards, Gompertz, logistic, monomolecular, Weibull and polynomials. In a posterior quantitative comparison, Gilligan (1990b) deals in some detail with the variability of curve parameters and also presents disease progress curves of some soilborne diseases which follow the monomolecular and the logistic functions. An example is presented in Fig. 5.15. A very detailed treatment to analyse and model the temporal aspects of the epidemiology of soilborne diseases is by Gilligan (1985, 1994).

In Fig. 5.15A, the effect of solarisation was associated with a reduced carrying capacity, the asymptote was affected, but not the locational and rate parameters and the time of onset. In Fig. 5.15B, the curves of five cultivars of raspberry did not show an effect on the rate parameter, but both the upper asymptote and the locational parameter varied (Gilligan 1990b).

Although temporal disease progress curves hardly differ between soilborne and foliar diseases, this may well be the case with their spatial behaviour (Chap. 6). Campbell and Benson (1994b) suggest that the spatial extent on dispersal patterns can be smaller than for airborne and foliar diseases and "still be ecologically meaningful and interpretable".

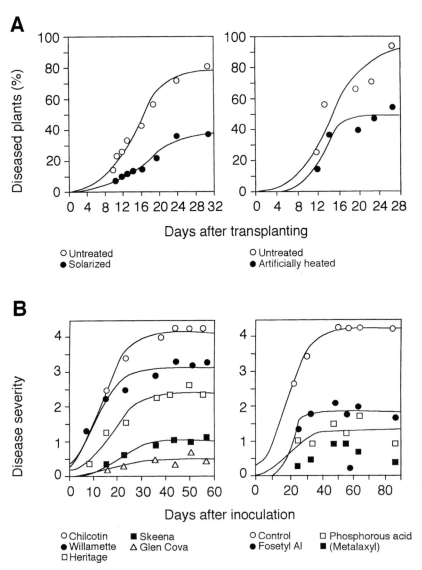

Fig. 5.15. Comparison of logistic disease progress curves of two soilborne diseases caused by **A** *Fusarium oxysporum* f. sp. *lycopersici* on tomato, untreated and treated and **B** *Phytophthora cryptogea* of raspberry on five cultivars. Data sources analysed by Gilligan 1990b (modified)

6 Comparison of Spatial Aspects of Epidemics

Spatial aspects of epidemics are mainly concerned with the local spread (e.g. within a field), invasion of pathogens (or their subpopulations) into other localities and factors affecting their persistence there; and, of course, ways and means of how to prevent invasion and persistence. There are excellent quantitative reviews of type 2 (Sect. 3.2.2.2) spatial and spatio-temporal aspects of plant diseases: Gregory (1968); Barker and Noe (1988); Gilligan (1985, 1988, 1990a, 1994); Fitt and McCartney (1986); Minogue (1986); Fitt et al. (1987); Jeger (1989, 1990); Madden and Hughes (1995a,b); Gottwald et al. (1998). Huber et al. (1998) and McCartney and Fitt (1998) reviewed specifically rain-splash dispersal. For soil-borne disease, Campbell and Benson (1994a) is recommended as a source of information. The reader should consult Carter (1973) for the spatial behaviour of viral diseases. Spatial interactions in ecology are treated in a book edited by Dieckmann et al. (2000), several chapters of which can inspire research on spatial aspects (and their comparison) in plant disease epidemiology, especially the chapter by Cox et al. (2000). However, spatial aspects of epidemics also vary stochastically, which is acknowledged, but "then widely ignored " (Gilligan 2002).

Studies across pathosystems on spatial aspects for comparative epidemiology are rare and the existing ones are mainly posterior analyses. Still, Hughes et al. (1997) feel that "It is debatable, however, whether there yet exists... a sound theoretical understanding of the range of observed spatial behaviour exhibited in plant disease epidemics. Although progress is being made, it is still hard to discern from the literature a broad consensus as to the way that the spatial dynamics of crop-pathogen systems can be summarized." This view points to a challenging task for comparative epidemiology to contribute to these ends.

6.1 Comparison of Spatial Distribution of Disease

Spatial distribution of disease is usually monitored on host plants. Functions as criteria for the comparison of the distribution in space are listed in Table 6.1. Zadoks and Schein (1979) distinguish in descriptive terms focal epidemics which start at low initial inoculum dispersed at random, regular, clumped, or aggregated in a confined area and general epidemics developing from high levels of well-dispersed inoculum. These authors list as spatial attributes the density as number of spores, lesions, etc. per unit area or space, disease severity and disease prevalence, i.e. diseased crops in relation to all crops of an area. These criteria are applicable to spatial patterns of disease at any scale that range from fields to large-scale surveys. The distribution patterns may be typical for the pathogen involved, based on its mechanisms of dispersal and spread of disease. The functions themselves (in Table 6.1) can be compared for their appropriate use in comparative epidemiology. Therefore, prior to planning comparative experiments, the most appropriate function for an analysis of spatial distribution may be tested first for a given problem. An introduction to spatial distributions and their implications, particularly for decision making, is given by Binns et al. (2000).

Table 6.1. Types of functions used in analysis of spatial patterns and their description. (Jeger 1990)

Statistical distribution	Indices of dispersion	Spatial autocorrelation
Binomial	Dispersion	Doublet analysis
Exponential	Cluster size	Ordinary run analysis
Gamma	Cluster frequency	General cross-product statistic
Geometric	Mean crowding	Moran's statistic
Negative binomial	Patchiness	
Poisson	Morisita's index	
Uniform (continuous)	Taylor's law	
Uniform (discrete)	Iwao regression technique	
Neyman type A		

For all the unrelated functions above see Jeger (1990, Table 3) for equations, parameters, means and variances

Some of the functions in Table 6.1, like the Poisson distribution, imply a random distribution. However, randomness is not the rule for diseases of crops. Among 112 distributions of disease that were compared, 98 (87.5%) were not fully random. There were 107 frequency distributions of lesions, including foliar infections, which best fit the negative binomial function (Waggoner and Rich 1981). The parameter k of the negative binomial indicates extreme aggregation when very small and a parameter value of k tending towards infinity stands for randomness. Gilligan (1988) refers to statistical tests for the deviation from randomness where two-dimensional lattices are involved. The doublet analysis and the ordinary run analysis are among the simplest techniques to detect non-random patterns. The results obtained may vary according to the level of disease incidence (Madden et al. 1982). Aggregation can be proven by these two analyses, for nearby plants or plant hills by the Greig-Smith index. The aggregation of disease within hills is determined by large variance-to-mean ratios and by the Morisita index (e.g. Johnson et al. 1991). Fifty epidemics of tobacco black shank (*Phytophthora parasitica* f. sp. *nicotianae*) and data from three locations in 1980 and one in 1981 were examined for the distribution of diseased plants. A non-random pattern was found in less than 10% of all cases in this study (Campbell et al. 1984). For the validation of mathematical models in plant-disease progress in space and time, Hughes et al. (1997) distinguished between methods of analysis for sparsely sampled and for intensively mapped disease incidences. For the sparsely sampled group, they used the binomial and beta-binomial distributions, the variance-to-mean relationship and the index of dispersion. Methods suitable for the analysis of intensively mapped sampling were the geometric-series test, ordinary runs analysis, spatial autocorrelation, geostatistics and distance- and distance-class analysis. Ferrandino (1998) proposed a radial correlation analysis to measure the probability of deviation from a random distribution based on a cumulative probability density function for pairs of diseased plants. A term r was introduced as an estimate for the scale of length along which disease was correlated.

Distribution functions found particular use for soilborne diseases. Attributes for the dispersal of soilborne diseases are (Campbell and Benson 1994a): inoculum source, dispersal, infection process and patterns and temporal stability. Spatial patterns of soilborne pathogens and associated disease are usually conserved and more recognisable over several periods of time, which may facilitate their comparison. Criteria which can be used for comparison are the variance-to-mean ratio (V/m), indices for mean crowding (m*), and patchiness (m~/m), and various frequency distributions (Poisson, negative binomial, Neyman type A etc.). However, comparisons by means of these functions fail to provide information on the spatial pattern of a disease in a field. For a comparison of spatial distributions, Campbell and Benson (1994b) refer to the nearest neighbour or the Greig-Smith methods and to spatial autocorrelation analysis. They also recommended the Moran's I coefficient which can range from −1 to +1. No autocorrelation exists if the value of the Moran's I coefficient for large sample sizes approaches 0. Where feasible, a two-dimensional spectral analysis could be employed (see Nicot et al. 1984).

Nicot et al. (1984) state that many techniques used in the analysis of spatial patterns do not take into account the relationships between locations of diseased plants in a field. Jeger (1990) accepts this criticism for frequency distribution analyses where information on locations is lost, but this criticism does not apply for doublet or run analyses. These techniques, by definition, when applied to continuous sequences, take both healthy and diseased plants into account. Their limitation of being one-dimensional only is overcome by the use of various forms of spatial autocorrelation analyses. For details of these techniques see Madden and Campbell (1990) and Campbell and Madden (1990).

Madden and Hughes (1995b) were especially interested in discrete distributions for binary data that are particularly applicable to disease incidence as the criterion to measure disease intensities. They included the binomial and beta-binomial distribution on which they elaborated in large detail. If the probability μ of a leaf, fruit, plant etc. being diseased or not is not constant, then the binomial distribution is not appropriate to represent the distribution of diseased entities in a sample. The reasons for a non-constant probability included variation in host, pathogen, and environment in a field, and factors that affect multiplication and dispersal. The distribution of diseased plants can rather be determined by mixing the binomial with the beta density function. This results in the beta-binomial distributions. For the derivation of this function and the index of aggregation, see Madden and Hughes (1995b).

Lesions on plants may have a random pattern which fits the Poisson distribution. However, frequently lesions have an aggregated pattern on diseased plants in a field that fits the negative binomial distribution well. For disease incidence, when binomial (random) or aggregated, the beta-binomial distribution had been established earlier as the more appropriate distribution (Hughes and Madden 1993,

Madden and Hughes 1995a). The latter was used by Laranjeira et al. (1998) on
Fatal Yellowing of oil palms in Brazil. For a citrus disease, the binomial distributions instead of the Poisson gave better approximations to the Taylor law
(Fig. 6.1). There was a good fit for this citrus disease in California and Spain,
where *Aphis gossypii* is the main vector. The pattern was interpreted as being in-
distinguishable from random patterns. Data from Costa Rica and the Dominican
Republic, where *Toxoptera citricada* is most important, were indicative of aggregation with the higher binomial variance.

In a proper across-pathosystem field experiment (p. 126), the spatial distributions of the incidence of four foliar diseases plus a complex of viruses were compared on white clover (*Trifolium repens*) and on tall fescue (*Festuca arundinacea*)

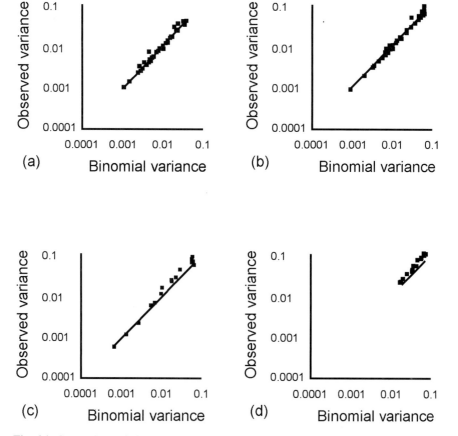

Fig. 6.1. Comparison of citrus Tristeza virus disease in groves of a California, b Spain, c
Costa Rica, and d the Dominican Republic. For each frequency distribution, the observed
mean and variance of disease incidence were calculated, as was the theoretical variance for
the binomial (random) distribution with the same means as that for the observed data. The
relationship between the observed (*dots* for at least 200 trees), and binomial variance (*solid
lines*) in the log-log co-ordinates. (Adapted from Gottwald et al 1998)

by Nelson and Campbell (1993b). The analyses were done with a two-dimensional distance class analysis for quantitative spatial attributes: viz. (1) number and size of clusters of diseased plants and (2) existence and magnitude of edge, row and column effects. The spatial patterns depended on whether the pathogens were splash or windborne. Significant edge effects and large, well-defined, expanding clusters were typical for *Pseudomonas andropogonis* and *Stagonospora melilotis*. Smaller and more loosely arranged clusters were formed with the wind-dispersed *Cercospora zebrina* and *Curvularia trifolii* that had shallower disease gradients. Viral diseases exhibited a stronger edge effect in the susceptible cultivar, but not so in the more resistant genotypes. Variability of spatial attributes between and within years, among plots and between vegetation periods suggest the importance of rain and temperature as well as harvesting practices as determinants of the spatial patterns.

Xu and Ridout (1998) proposed a stochastic model to simulate the spread of disease in space and time. They used this model to study the effects of number of initial inocula and their spatial patterns, sporulation rate and spore dispersal gradient on the spatio-temporal dynamics of plant disease epidemics with the Cauchy distribution with "median dispersal distance". A similar term is the "probable flight distance" (Table 4.13) by Schrödter (1960) where 50% of the propagules are lost from the spore cloud. This dispersal distance may also depend on a seasonal availability of a susceptible host to ensure coincidence with a virulent pathogen and favourable weather conditions to cause epidemics. In their simulation, Xu and Ridout (1998) found that the rate of temporal increase was mainly influenced by the sporulation rate (spores per lesion per day, i.e. the sporulation intensity).

6.2 Foci and Gradients of Dispersal and Disease

6.2.1 Foci

The spread of diseases starts and develops in foci through the dispersal of inoculum. Foci are local concentrations of disease (at a time t) becoming apparent when disease multiplies (Van der Plank 1963). Growth of an individual focus varies in form and steepness of gradients which depend on the mechanisms of propagule dispersal of pathogens, distribution of susceptible plants, topography and, beyond the individual field, on cropping patterns of host plants. Inoculum from a focus by dispersal produces daughter foci and increases the population of foci in a field and beyond. Foci and their disease gradients may be rather inconspicuous at the beginning of an epidemic before they enlarge. As foci grow, infection mainly spreads from the perimeter where diseased and healthy plants meet, and the rate of growth of the focus slows down as the focus enlarges, other things being equal (Van der Plank 1963). However, a constant rate is often assumed in models (see below). New and growing foci sooner or later merge and disease is then said to have "generalised" in a field. If there are numerous foci at time t_0 of primary disease intensity y_0, the disease seems to start uniformly (and somewhat explosively) without distinct gradients. Many initial foci commonly occur with seedborne pathogens (Baker and Smith 1966) or with diseased transplants (Berger 1973).

Criteria for comparisons across pathosystems are the focus size and its expansion rate, which is correlated with the daily multiplication factor (i.e. the product of spore production and infection efficiency, also known as the progeny/parent ratio). The expansion is inversely related to the latent period, that is, a long latent period slows focal expansion (Hau and de Vallavieille-Pope 1998). The radial speed of focus expansion was assessed to be about 10 cm/day for stripe rust (van den Bosch et al. 1988b), 9.6–61.5 cm/day for leaf rust (Subha Rao et al. 1990) and about 20 cm/day for stem rust (Schmitt et al. 1959). As descriptive terms Zadoks (2000) classified foci in three epidemic orders called zero-, first-, and second-order epidemics depending on the criteria size, complexity and time scale. In principle, all three orders of focal epidemics obey the same general rule, that of "constant radial expansion". This, of course, is a theoretical assumption. Zero-order epidemics are small-scale foci of about 1 m in diameter. These epidemics are common in many well-known pathosystems. First-order foci after four to six generations of dispersal remain small (max. up to 20 m in diameter). With favourable weather conditions, these primary foci enlarge and daughter foci appear in the same and neighbouring fields. Their size may be restricted to a field or attain any size up to 1000 km in diameter. Second-order epidemics may be pandemics. Zadoks lists examples from the literature for the three above classes he defined.

For focal expansion, two theories for the dispersal of inoculum were proposed by van den Bosch at al. (1988a): a rate of spread that asymptotically approached a constant value. The calculations with their models were based on the concept of a time kernel (i.e. inoculum production through time), contact dissemination and gross reproduction (i.e. total number of infected plants produced by a single infectant). Time kernel and the spatial contact distribution of daughter lesions around a mother lesion were fitted to experimental data of *Puccinia striiformis* on wheat and *Peronospora farinosa* on spinach (van den Bosch et al. 1988c). For both pathosystems, the observed and expected rates were in good agreement. The accuracy of the estimated variance of the contact distribution has a large influence on the accuracy of the predicted velocity, particularly when gross reproduction is small. Sufficiently flexible subclasses in the models can easily be adapted to various pathosystems (van den Bosch et al. 1988b). One subclass was a mechanistic submodel for dispersal of inoculum based on turbulent diffusion inside the canopy and with random interception of this inoculum by host plants. The inoculum production is described by a shifted gamma density distribution function (Table 6.1). Contact distribution in this model follows a concept of primary gradients of a focus, the velocity of focus expansion and its front. Constant rates, as used in this model, are common in a number of pathosystems. Contact between healthy and diseased roots may be important in soilborne diseases. Pathosystems known to have focal expansion at a constant rate may change their patterns when environmental conditions change, or, for instance, when a vector becomes involved. Then increased rates of focal expansion are rather the rule as described by Ferrandino (1993; see p. 152). It would be a challenge for comparative epidemiology to establish which of the two types of focus expansion prevails for pathosystems under the conditions in which they occur.

For a comparison of the spatial dynamics of epidemics, Jeger (1985) sees three areas where simple mathematical models may prove useful. That is "(1) the generalization of disease gradient models into models of increase in time and space by considering the rate of isopathic movement from an initial focus, (2) the use of random walk theory to consider the expansion of disease along the rows of agricultural crops, and (3) the use of characteristic spacing of many crops in regular lattice, and the implications of this regularity for modeling." More descriptive criteria to compare foci are steep or flat gradients, the typical form, whether edges are sharp or diffuse, and the tendency to form outward daughter foci.

6.2.2 Gradients

Gradients (Figs. 6.2 and 6.3) are the graphs of spatial dynamics in epidemics inside individual foci, and beyond by means of daughter foci. They express the spatial dynamics of pathogens (dispersal gradients) and diseases (disease gradients). Criteria for gradients are: distance, spore catches (e.g. number of spores/h/m^3), disease intensity, a degressive leg, an inverse asymptote and the isopathic rate (see p. 146). Observed dispersal gradients of three diseases are presented in Fig. 6.2. The relationship between disease gradients and temporal disease progress is shown in Fig. 6.3.

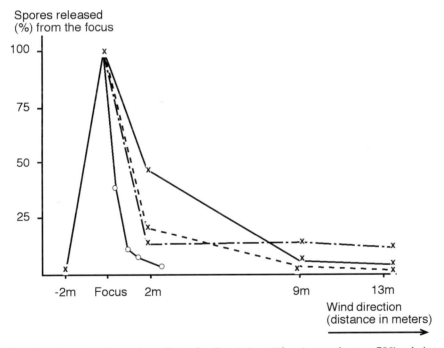

Fig. 6.2. Observed dispersal gradients for *Puccinia striiformis* on wheat at 70% relative humidity (*x, continuous line*) and 90% relative humidity (*x, dashed line,*), *Pyricularia grisea* on rice (*open circles, continuous line*) and *Erysiphe graminis* on wheat (*x, dashed–dotted line*). (Rapilly et al. 1970; Notteghem 1977)

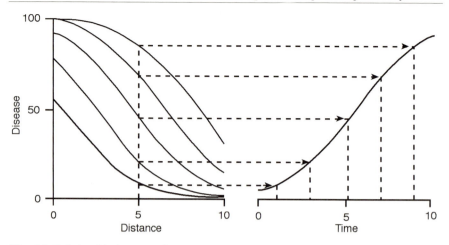

Fig. 6.3. Relationship between disease gradients and disease progress curves. On the *left*, disease readings are taken at a fixed point as the wave of disease passes at constant velocity. When the state values are plotted over time (*right*) the disease progress curve has the same functional forms (here: logistic) as the gradient. (Minogue 1986, with permission)

Dispersal gradients are affected by the shape of the source of inoculum (point, line, strip, area), height of inoculum source above ground, its intensity of inoculum production (Gregory 1968) and the mechanisms of liberation (e.g. windborne, splashborne, vectors). Under the same conditions, more potent sources of inoculum over time produce flatter and more far-reaching gradients, although the bulk of the inoculum always lands near the source. Pathogens that are windborne and pathogens dispersed by winged vectors tend to have flatter gradients with rather diffuse margins. Splashborne inoculum (bacteria or fungal spores by water droplets) and viruses by leaf contact or by vectors of limited mobility have steep gradients. The steeper the gradients, then the sharper the margins of the foci will be (Van der Plank 1963).

Dispersal gradients can be described by a variety of empirical functions (Wolfenbarger 1959; Gregory 1961, 1973; Minogue 1986; McCartney and Fitt 1998). Here we follow Campbell and Madden (1990). The (negative) exponential law (or log-linear) function is $y(x)=a \exp(-bx)$. The a is a constant, e.g. $y(0)$, the source strength measured as the amount of inoculum or disease intensity at the source), b the rate of decrease in y with the distance x from the source. The function is linearised as $\ln(y)=\ln(a)-bx$. For the (inverse) power law (or log-log), the function is (with the same terms as in the exponential) $y=ax^{-b}$, This latter function is linearised as $\ln(y)=\ln(a)-b \ln(x)$. The two functions are compared in Fig. 6.4. The constant a here is $y(1)$ for either intensity measured at the first point of measurement 1 (in cm, m etc.). Both functions are descriptive only, but useful for comparisons. Hence, the function that best fits the observed data should be preferred. However, there are indications that the power law tends to overestimate deposition near the source, while the exponential law may underestimate it (Fitt and McCartney 1986). The exponential function may be more suitable for splashborne inoculum, the power law function for small windborne spores. Although the two functions in Fig. 6.4 may have similar shapes their fundamental difference is that the exponential equation implies a constant length scale d (in which the density of spores etc. c, which is related to the strength of the source, decreases by the same proportion over equal distances), while the inverse power function implies a length scale that changes in proportion to distance. A fixed length scale for the power law leads to the concept of the half-distance $d_{1/2}$ or α (=ln2/d or 0.693/d), the distance in which c decreases by half, as an easy method to visualise gradients. For more details on the application of either function and their relative advantages see Minogue (1986), Fitt and McCartney (1986 with an update by McCartney and Fitt 1998), Campbell and Madden (1990) and Jeger (1990).

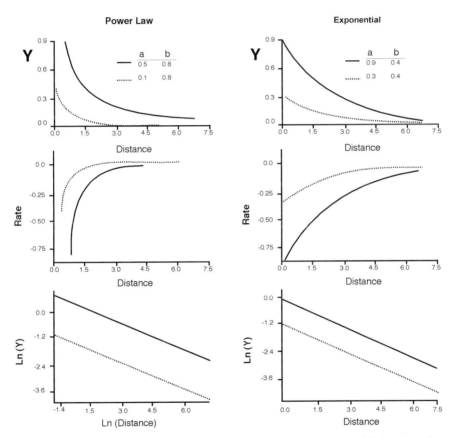

Fig. 6.4. Hypothetical dispersal gradient by the power law and exponential functions (see equations in text) and the rate of change of y with the distance from the source and in their linearised form. (Campbell and Madden 1990, with permission)

Equations for the deposition of spores that emanate from point, line, strip, and area sources are summarised by Gregory (1961, 1973). Gregory (1968) also compared 124 dispersal gradients of dry airborne fungi and pollen. Among the gradients included in the evaluation, 59 gradients were fitted better by the power law and 65 by the exponential law functions according to r^2 as the test parameter for the linear regressions. An advantage of the power law model is that the exponent b (or β) is independent of the units in which distance is measured, whereas d in the exponential law is not. For this reason, the power law function apparently copes better with large differences in the length of scale.

Fitt and McCartney (1986) referred to Gregory (1968) and concluded that for the dispersal of dry airborne spores, their gradients may fit both functions equally well. Differences nevertheless exist, e.g. the splashborne gradients of *Pseudocercosporella herpotrichoides* in both still and moving air fit the exponential law better . Rapilly (1991) compiled some observed gradients of pathogens that fit the power and exponential functions (Table 6.2). Both functions had high coefficients of determination (r^2) for all the pathogens compared, except for *Erysiphe graminis*. The dispersal gradients of the three rust fungi can be described equally well by the power and exponential laws. The half-distance of the dispersal gradient, which is independent from the point from which it is measured, was between 2.4 and 2.7 m for *P. striiformis*, 5.8 m for *E. graminis*, 25 m for *P. graminis* and 28 m for *P. recondita*.

Table 6.2. Functions to approximate observed dispersal gradients with the power and exponential law. (from Rapilly 1991, his values rounded)

Pathogen	Power $y=ax^{-b}$ $y=$	r^2 $y=$	Exponential $y=a\exp(-dx)$	r^2
Puccinia graminis	$91.99x^{-0.026}$	0.81	$102.02e^{-0.028x}$	0.91
P. recondita	$96.96x^{-0.022}$	0.93	$98.91e^{-0.025x}$	0.98
P. striiformis at 90% relative humidity	$106.83x^{-1.54}$	0.99	$61.14e^{-0.29x}$	0.94
P. striiformis at 70% relative humidity	$134.40x^{-1.29}$	0.95	$90.67e^{-0.26x}$	0.99
Erysiphe graminis	$68.26x^{-0.74}$	0.79	$46.08e^{-012x}$	0.55
Pyricularia grisea	$105.45x^{-2.82}$	0.99	$75.21e^{-1.40x}$	0.94
Alternaria brassicicola before harvest	$333.38x^{-0.89}$	0.81	$35.30e^{-0.0056x}$	0.74
Alternaria brassicicola at harvest	$329.45x^{-0.66}$	0.71	$57.61e^{-0.0028x}$	0.87
Uromyces appendiculatus	$182.75x^{-1.93}$	0.77	$150.52e^{-0.71x}$	0.93

For the power law function x+1 was used, y is in %

In a quantitative posterior analysis and comparison, Ferrandino (1996) applied linear regression of the exponential and power law models to compare 105 dispersal gradients ("profiles") obtained from the literature. Fifty-one of the data sets fit the exponential law better and 54, the power law functions, based on a simple comparison of the coefficients of determination (r^2). However, when tested with the F-test, only 6 of 51 that favoured the exponential law and 15 of 54 cases that favoured the power law were significant at $P<0.05$. In general, experiments with a point source of inoculum tended to fit the exponential law and experiments with a line source of inoculum favoured the power law. The comparisons above show little preference for either empirical model, as 80% of the gradients were equally well described by either function. This agrees with Fitt et al. (1987).

There is good agreement among authors (e.g. Berger and Luke 1979; Minogue 1986) that, in general, the log-linear transformation for the exponential law is probably as successful as the log-log transformation for the power law in straightening gradients. However, Minogue (1986) favours the latter because of its superior mathematical and interpretational qualities and recommends its use whenever appropriate. Neither of these two functions makes any allowance for multiple infection. Even if corrected, the functions are, according to Minogue, only applicable for primary gradients or secondary gradients early in the epidemic. Their use is limited to follow gradient development throughout epidemics. Berger and Luke (1979), however, feel that this limitation may be overcome by the logit-log transformation.

Dispersal of soilborne pathogens is compared in detail by Campbell and Benson (1994b) who cite numerous experimental results. The major difference between foliar and soilborne diseases lies in the fact that the propagules of the inoculum of the latter commonly are "formed and released within the solid matrix of soil" (Campbell and Benson 1994b). Consequently, propagules of soilborne diseases are dispersed only a very short distance below ground. In the soil matrix, the movement of propagules of soilborne pathogens depends on a number of physical factors including moisture, temperature, particle size distribution and aggregation, soil texture, and channels produced by animals, insects or decaying roots.

For soilborne pathogens, except for some studies with *Phytophthora* spp., little is known about these interacting factors (see Campbell and Benson 1994b for details). This implies that propagules are dispersed only short distances below ground. Many pathogens have mechanisms, like fruiting bodies (e.g. *Sclerotinia* spp., *Gaeumannomyces graminis*), to liberate spores above ground for wind- or waterborne dispersal. Propagules may also be moved out of the soil matrix, become windborne and be dispersed for long distances by dust, aerosols, impacting rain or plant debris. Contaminated irrigation water can introduce soilborne pathogens into fields. This type of dispersal is well documented for several *Phytophthora* spp. and *Pythium* spp. Dispersal of pathogens also occurs with shipments of plants with soil on their roots, cultivation equipment, boots, tires of farm machinery and hooves and skins of roaming animals (Campbell and Benson 1994b).

Disease gradients are the result of dispersal gradients of pathogens along which they have been effective in causing disease. Epidemiologically, disease gradients are more relevant and depend on the following factors, viz. the means of dispersal, viability of inoculum at landing, environmental factors that favour infection and density and distribution of susceptible host plants. The two essential parameters for disease gradients are distance from a known or assumed source d (or x), and disease intensity y along a defined path (the gradient). When y is plotted against d or x on the abscissa, a hyperbolic curve 1/y will be obtained. This leads to the biological disease gradient. Other factors like light and exposure to, or protection from, wind (e.g. by hedges) cause physical gradients of disease (Fig. 6.5).

In the very dry summer of 1976, a distinct physical gradient of powdery mildew in spring barley with high disease severity (BS) was recorded within the

range of an extended morning shadow from a forest along one side of the field
(Fig. 6.5). In the following wet summer, it was the reverse, i.e. less disease intensity near the forest than in the open field. A crop in the neighbourhood with overseasoning inoculum or alternate hosts can give rise to biological gradients. A field
of spring barley sown adjacent to a crop of winter barley with overwintering powdery mildew (*Erysiphe graminis*) had a distinct biological gradient of disease severity early in the season before disease had generalised in the field (Koch 1980).

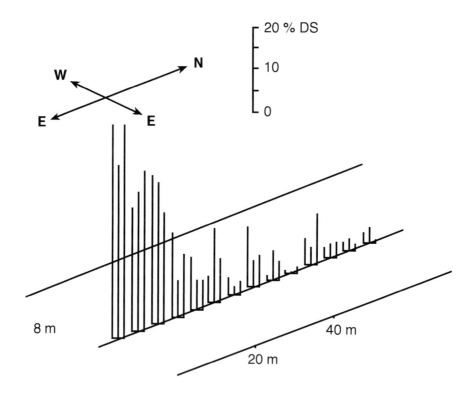

Fig. 6.5. Physical disease gradient of barley powdery mildew severity (indicated as *bars*, in
% disease severity DS) within the area of long-lasting morning shadow from a forest of tall
trees. (Koch 1980)

The spatial spread of disease along its gradient follows a "rate of isopathic
movements" (Fig. 6.6), a concept introduced by Berger and Luke (1979; Table 6.3). An isopath connects points of equal disease intensities. The rates of isopathic movements proved useful to compare the spread of disease more appropriately and precisely. Isopaths are related linearly to the distance from the source of
inoculum. Isopathic velocities, i.e. their rates through space, depend on prevailing
disease intensities, time and distance, either with constant, decreasing or increasing rates.

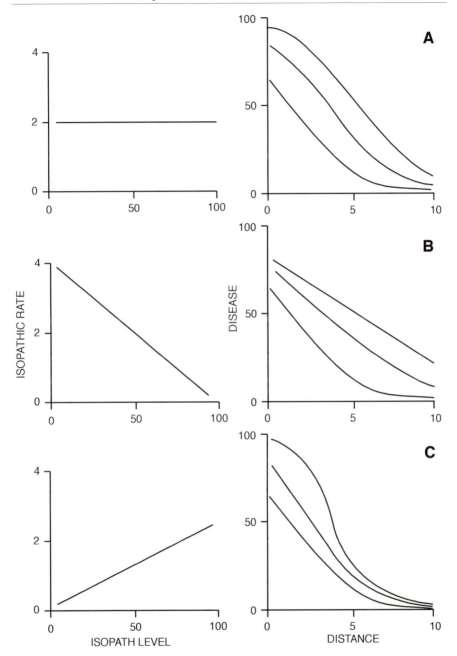

Fig. 6.6A–C. Effect of isopathic rates (*left*) on the development of the gradient (*right*). Depending on whether the rate of movement of the isopath is A constant, B decreases or C increases with disease intensity, the gradient, respectively, remains stable, flattens or becomes steeper with time. (Minogue 1986, with permission)

Table 6.3. Velocities of spread of several diseases. (Compiled from various sources by Minogue 1986)

Pathogen	Host	Velocity
Pseudocercosporella herpotrichoides	Wheat	0.2–0.3 m/day
Stagonospora nodorum	Wheat	0.3 m/day
Puccinia coronata	Oats	0.2–1.2 m/day
Phytophthora infestans	Potato	3–4 m/day
Ceratocystis wageneri	Pine	0.1–1.8 m/year
Tomato ring spot virus	Raspberry	2 m/year

In a model computation with analytical functions for the three-dimensional and turbulent dispersal of airborne spores, a contact distribution was found which increased by a length scale in the downwind distance (Ferrandino 1993). Such contact distribution approached an inverse power law at long distances. Simulated epidemics based on this function exhibited spatial disease gradients that became shallower as the epidemic progressed. If the spatial co-ordination is log-transformed first, than the leading edge of this dispersive epidemic wave propagated more quickly than did the trailing edge (Ferrandino 1993). As a result, the wave spreads outward in space with increasing time. This behaviour contrasted the constant isopathic velocities characteristic of the travelling wave description as predicted by spatial contact distributions of an exponential order that had a bounded length scale.

Additional details about disease gradients of mono- and polycyclic diseases are treated by Minogue (1986). He also identified "outstanding questions" in focal spread of disease: the spread in two dimensions, non-wavelike spread, the effect of plant density and spatio-temporal relationships, to which scaling effects may be added. For the comparison of two slopes of gradients, he proposed the "gradient equality" with measurements at equivalent points on both gradients permitted; "fixed-x" (on the x-axis) for space and "fixed-y" for disease progress. Gradients are considered to be equal if their derivates $dy/dx=-by$ are the same when measured at equal levels of disease. This is the only valid measure for gradient steepness, if the equality of the parameter implies equality of gradients in this sense (Fig. 6.7).

The local spread of rugose mosaic virus (RMV) and potato leaf roll virus (PLRV) was compared in an experiment over 3 years at several sites (Gregory and Read 1949, cited by Thresh 1976). Over a distance of about 2 m, the gradients differed quantitatively in all 3 years because the rugose mosaic virus caused less infection. The shapes of the gradients for both diseases were similar, although one gradient of RMV dropped somewhat steeper at about half the distance than the gradient for PLRV. There was essentially no difference in these gradients for virus diseases compared to the dispersal gradients of fungal pathogens. Similarly, the proximity of older or overwintering crops with abundant inoculum ensures heavier and earlier infection with obvious disease gradients away from the inoculum sources. Particularly well-adapted pathogen genotypes from these adjacent sources can have a pronounced effect on disease gradients. When experiments are conducted under such conditions, it should be determined whether these external host

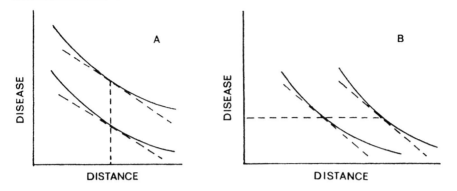

Fig. 6.7. Gradient steepness measured as the derivate (*dotted line*) A at a fixed distance or B at a fixed disease intensity. (Minogue 1986, with permission)

factors may have influenced the gradients. Comparison of these aspects will be mainly descriptive.

For comparative epidemiology on the community level, practically the same criteria apply. However, the problem of scale (distances) and the distribution pattern of susceptible crops have to be observed when communal aspects are compared (Zadoks and Schein 1980). Proper descriptive and measurement terms to compare density and size of fields with susceptible crops still have to be proposed. For the time being, available information on distances and acreages may be used as measurement terms.

6.3 Long-Distance Spread

Many diseases have spread worldwide by long-distance dispersal of their pathogens to areas wherever vulnerable host plants grow and favourable environmental conditions prevail. This long-distance spread (Sect. 4.3.2.5) is facilitated by air currents, birds, floating plants, seeds and various human activities (e.g. trade, tourism). Studies on the long-distance spread of disease and the dispersal of pathogens have been treated in case studies, descriptions of trajectories and theoretical considerations by Schrödter (1960), Gregory (1961, 1973), Hirst et al. (1963), Zadoks (1967), Nagarajan et al. (1976, 1984), Nagarajan and Ajai (1988), Aylor et al. (1982), Aylor (1986) and Pedgley (1982, 1986). However, consolidated and generally accepted theories of long-distance spread have not yet emerged.

Descriptive and measurement terms for a comparison of trajectories in long-distance dispersal are: inoculum sources, conditions which favour its liberation and lift into higher air strata, direction of wind currents (hurricanes and cyclones included), factors that act upon inoculum deposition by sedimentation or wash-out by rain and, finally, the survivability of infective propagules. All this has to coincide with susceptible stages of the crop that receive the inoculum. The identification of sources, e.g. somewhere in Turkey, and the monitoring of landing inoculum of *Puccinia striiformis* and *P. recondita* in northern India was crucial (Nagarajan et al. 1982). The travelling spore cloud of *Puccinia* spp. could be tracked with weather satellites by the cloud drift of a rain front. Satellites and GIS methods will be useful tools to link all the above-mentioned aspects of long-range dispersal and these tools will assist the development and validation of predictive models. For the identification of inoculum sources, molecular methods like RAPD may help in the future.

Dispersal may happen in waves or in a single jump. More recent examples for a spread in waves – "a series of micro- and mesoscale events" (Campbell and Madden 1990) are the black Sigatoka disease of banana (*Mycosphaerella fijiensis*), tobacco blue mould (*Peronospora tabacina*), sugarcane rust (*Puccinia melanocephala*), sugarcane smut (*Ustilago scitaminea*), coffee leaf rust (*Hemileia vastatrix*), barley yellow dwarf virus etc., causing severe initial losses typical for pandemics. These waves usually occur over a number of years (for details see Aylor 1986). A one-jump spread was observed with *Melampsora medusae* and *M. larici-populina* when these rusts travelled from Australia to New Zealand (Kraayenoord et al. 1974). Similarly, the tropical rust of maize (*Puccinia polysora*) may have been introduced from the Americas to West Africa, or the chestnut blight (*Cryphonectria (Endothia) parasitica*) from the Far East or Europe into the United States. Seasonal spread of pathogens across continents is common, e.g. wheat stem rust (*Puccinia graminis*) in the United States, from North Africa to Europe. Wheat stripe rust (*P. striiformis*) spread from southern India to the Himalayan area.

Schrödter (1960) advanced functions with measurement terms like the ones in Table 6.4 (A, c_e, and u, see also Table 4.13) to estimate probable flight range and duration of spores based on studies on atomic fall-out and dust transport. For the estimated distance of dispersal, he used the Schmidt-Rombakis method and Sutton's equation in the concentration of the spore cloud with spores of four sizes expressed as c_e in Table 6.4. For the modelling of long-distance dispersal, Aylor (1986) and Davis (1987) suggested eight aspects: (1) inoculum source characteristics; (2) advection to facilitate transport from the source; (3) horizontal and vertical dispersion of spore clouds; (4) changes in the composition of travelling clouds; (5) loss (and gains) of inoculum in air parcels; (6) dry deposition; (7) wet deposition and (8) fate of spores on host surfaces.

Table 6.4. Probable flight duration of spores of different size. (Schrödter 1987)

A^a	c_e^b in still air (cm/s)			
(g/cm/s)	0.035	0.138	0.975	1.300
10	72 days	4.6 days	2.2 h	1.2 h
20	144 days	9.2 days	4.4 h	2.5 h
50	360 days	23 days	11 h	6.2 h

[a]The air mass exchange coefficient; here.
[b]The fallout speed of elliptic spores. No wind speed u as in Table 4.13.

Long-distance spread, apart from the introduction of new pathogens to an area and the concerns of plant quarantine, carries along additional inoculum. Also, new races of a pathogen may be spread to infect a cultivar that hitherto was resistant to this species of pathogen. Limpert (1987) estimated the downwind movement of races of barley powdery mildew in Germany at about 110 km/year, probably in multiple jumps from intermittently infected fields. In pathosystems with a gene-for-gene relationship, migration of pathotypes and their genes occurs as gene flow, i.e. the movement of genes to other pathogen populations. At first, this may be cryptic before being locally subjected to selection. Gene flow in barley powdery mildew, Septoria leaf blotch of wheat, potato late blight, rice blast and stem rust of wheat have been described in detail (McDermott and McDonald 1993). However, Sun and Zeng (1995) concluded from model computation and their own experiments that with a disease intensity of y >0.05 in a recipient region, the contribution of spores from one region to disease development in another region will be very small. The amount of spores that arrive may be further reduced by the loss of vitality during transport over long distances. Campbell and Madden (1990) quote short "jumps" of *P. graminis* at 54 km/day in the nearly continuous wheat areas from Texas to Minnesota in 1983. The longest well-documented "jump" of urediospores of wheat stem rust in North America by Roelfs (1985) was about 680 km between two wheat growing regions in Canada separated by forests and lakes (Campbell and Madden 1990). Nagarajan et al. (1976) postulated a distance of 970 km covered by urediospores of *P. graminis* in India.

From the geobotanical point of view, long distance dispersal may bring about new encounters and re-encounters of pathogens and hosts (Gäumann 1951). A new encounter happens when a plant species is introduced into an area where it had not existed before and encountered an endemic micro-organism that then turns into its new pathogen. A pathogen may be dispersed and find a new host plant in another region of the world. Re-encounters take place when a pathogen again meets its host from which it was separated by man, or due to phylogenetic reasons. Gäumann (1951) cites, among others, *Uncinula necator* on *Vitis vinifera* was transported from America to Europe, and *Cryphonectria (Endothia) parasitica* on *Castenea* spp. from Europe to America. Robinson (1976) gives an extensive account on the re-encounter of *Puccinia polysora* left behind in the Americas with its maize host in West Africa after nearly 400 years. A case where a re-encounter has been obviated by quarantine and breeding for resistance is the South American leaf blight (SALB) of rubber (*Microcyclus ulei*), which still is confined to South America.

Experimental comparison of long-distance dispersal admittedly meets with great difficulties. For the time being, analysis and comparison of trajectories of long-distance inoculum transport still appear somewhat erratic, in spite of empirical evidence that inoculum can travel over large stretches (see Gregory 1961, 1973, for example). Apart from technical difficulties, no unified concepts for an investigation of long-distance dispersal have yet been agreed upon. Therefore, specialists should embark upon posterior analyses to develop some common concepts or consolidated theories.

6.4 Comparison of Geophytopathological Aspects

As stated before, the invasion and persistence of pathogens in a field, region, or globally, are essential issues of spatial aspects of epidemics. Park et al. (2001) pointed out that studies of invasion naturally need a concept of invasion thresholds which are often related to the basic reproductive function iR (Van der Plank 1963) or R_0, usually defined as the average number of new infections produced when a single propagule is introduced into a wholly susceptible host population. For a simple model with logistic host growth and a polycyclic disease, invasion results from the product of the transmission rate multiplied by the equilibrium density of the susceptible hosts and the infectious period of the disease. An epidemic will then occur when iR>1. They consider such a model too restrictive and suggested some realistic criteria, primary amongst which are:
– The assumption that invasion begins when the host is at equilibrium;
– The models assume that there is simple mixing between infected and susceptible hosts;
– There is no allowance for host growth;
– There is no allowance for spatial heterogeneity and the interactions that occur between adjacent fields in sharing inoculum; and
– The models are deterministic, so they do not tell us about risk.

Such models can be expanded by allowing secondary infections in addition to immigration. Also, the changes in host response to infection as well as changes in transmission terms (e.g. mixed host populations) restrict transmission, which may cause more complicated thresholds for invasion, e.g. the threshold population of susceptible and infected host plants that are necessary for invasion. Criteria for persistence are a threshold reservoir of inoculum, possible co-existence of disease and host plant(s), little or no disturbance factors such as harvest or the onset of unfavourable conditions for disease and the risk of fading-out during saprophytic stages or hostless periods.

Experimental comparison at the community level and for agro-ecological regions is, by far, less developed than that of epidemics in a field. One such attempt was made in a study on the parasitic mycoflora in six vegetation types at each of four sites with different soil types for 3 years (Kranz and Knapp 1973). Monthly, from April to October, all identifiable pathosystems were monitored in each plant community. The criteria used in this study for 3 years were: (1) host plant species; (2) host growth stage; (3) pathogen species; (4) disease incidence and severity at each of the 7-monthly assessment dates.

Venn diagrammes (Fig. 6.8) were used to make visual comparisons of commonalities and differences of pathogen species in three vegetation types. These types also differed among three localities with different soil types with the initials of the localities being D, N, W. The fungal taxa of pathogens that occurred in the vegetation types and localities are coded with numbers. The clear names of the fungal species (see Kranz and Knapp 1973) are irrelevant in this context. What is essential is which taxa (numbers) occur in which vegetation types and localities.

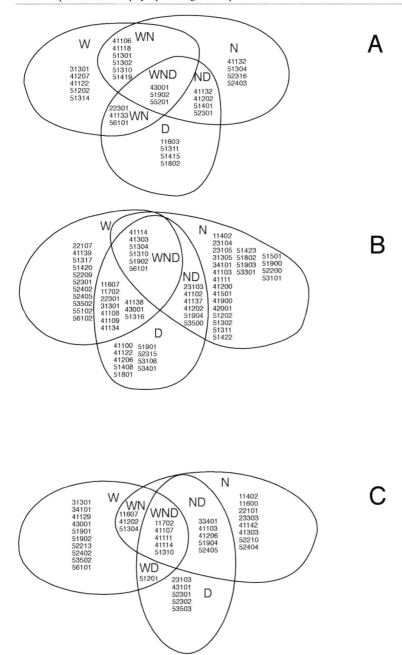

Fig. 6.8A–C. Venn diagrammes with pathogens (code numbers: *1* for Peronosporales, *2* and *3* for Ascomycotina, *4* for Teliomycetes and *5* for Deuteromycotina) existing in three plant communities (A forest, B dry meadow, C meadow; for the three localities *D, N, W*). Similarities in pathogens species are described by species in the combinations (e.g. *WN*) and differences by those outside the intersections. (Kranz and Knapp 1973)

   The three vegetation types in Fig. 6.8A–C differ first in the number of pathogen species that occurred, but also in the composition of the mycoflora in the vegetation types A, B, and C at three locations with different soil types. In a discriminant analysis with data from these experiments, fungal pathogens were statistically weighted and sites and vegetation (plant associations) characterised by assigning taxa to them (Fig. 3.6). From these analyses, one could anticipate the proneness for pathogens to occur in an area and determine the risk.

   Weltzien (1972) started with geophytopathological studies on a geographically larger scale. As data bases he used his own appropriately designed surveys, reliable reports of disease occurrence (or prevalence) and intensity, country or host lists and maps (like the CMI distribution maps), and climatic diagrammes to compare the distribution and regional importance of pathogens and diseases (Weltzien 1988). From data and information obtained with these approaches, techniques and his own surveys on disease intensity and losses incurred, Weltzien (1972) recognised three types of disease proneness:

1. areas of main damage, where epidemics occur whenever a susceptible crop is grown without protection;
2. areas of marginal damage, where epidemics occur irregularly and cause significant damage in some seasons only;
3. areas of sporadic attack, where damaging disease rarely occurs.

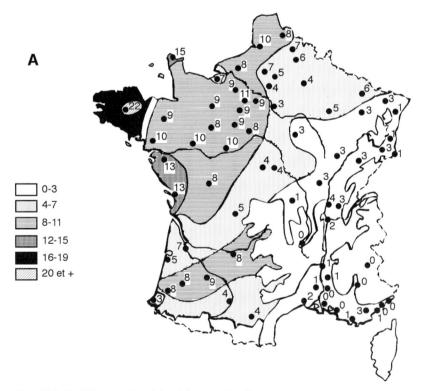

Fig. 6.9A, B. Mapping the risk of favourable climatic factors conducive to infection of wheat by *Pseudocercosporella herpotrichoides* and sunflower by *Sclerotinia sclerotiorum* in France. A Cases of eyespot between 16 November and 31 January; B *Sclerotinia* attacks (0 and above 10%) during August. Numbers in the maps refer to records received. (Rapilly 1991, with permission)

The evaluation of satellite imageries (Nagarajan and Ajai 1988) and the potent techniques of the Geographical Information System (GIS, see Nelson et al. 1994) will assist more quantitative large-scale comparisons in the future. Nelson et al. (1999) demonstrated the potential of various applications of GIS on the spatial distribution of plant diseases for comparison and plant disease management programmes. Teamwork, however, is necessary to cope with costs and time needed for the use of GIS and geostatistics, particularly when comparisons of dynamic spatio-temporal aspects are planned.

Rapilly (1991) identified and mapped zones of varying risks for the most important diseases based on information from the national weather service. Two pathosystems, *Pseudocercosporella herpotrichoides* on wheat and *Sclerotinia sclerotiorum* on sunflower, are shown as examples (Fig. 6.9A,B). The ratings on the maps are based on accepted meteorological conditions favourable for infection and the frequencies of their occurrence. For *P. herpotrichoides* (Fig. 6.9A), the risk of occurrence is, for instance, eight- to nine-fold higher in the Basin of Paris than in the east of France. Head blight of sunflowers (Fig. 6.9B) has a zero risk in northern and western France, but constitutes a high risk in the south. Maps of this type are useful tools for geophytopathological comparison. The data for the index values and entries in the maps are from 25 years for the wheat and 50 years for the sunflower pathosystem. Their usefulness for forecasting and disease management is obvious.

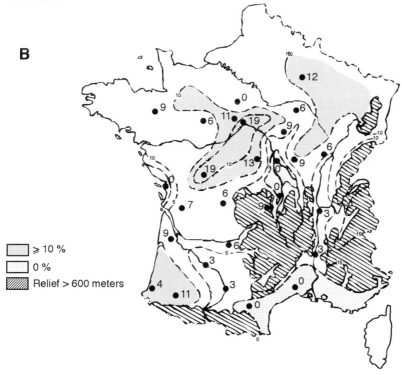

Fig. 6.9 B

An outstanding example of comparative epidemiology on an international scale is the one on witches' broom of cacao (*Crinipellis perniciosa*) in South America. Rudgard et al. (1993) first described the methodology and the site profiles of a joint project conducted in countries where the disease occurs on cacao. Six country reports following the same format presenting results on comparative epidemiology of the disease are summarised for relevant aspects of the epidemic by Schmidt et al. (1993).

7 Comparison of the Effects of Epidemics

One of the ultimate objectives of research in plant disease epidemiology is the development of supportive measures in decision-making for more effective, rational and less hazardous disease management in systems control and design. Comparative epidemiology could also play a role in the validation of proposed solutions and improvements using criteria and procedures described in the preceding chapters. For example, Binns et al. (2000) elaborate on comparisons of sampling plans to enhance and evaluate their usefulness. This includes the effects that diseases have on crops (Sect. 7.1 and 7.2) as well as the efficiency of control measures (Sect. 7.3).

7.1 Disease Intensity – Yield Loss Relationships

Disease, according to Burdon (1993), may cause damage to plants as castrators (e.g. some smuts, some rusts, choke in systemic diseases), killers (e.g. vascular wilts, damping-off) and debilitators (e.g. viruses, many rusts, mildews, leaf spots). Debilitators certainly interfere negatively with the source-sink relationship of the host plants that cause yield loss as accumulated photosynthates in crops. If the affected organ is the unit of yield, e.g. tobacco leaves or fruit, loss is proportional to the number of organs with disease. Blemished fruit on the market may downgrade the quality and thus the price of the product, which then incurs a loss of income. However, some plants can tolerate a certain amount of disease without measurable loss in yield, i.e. they have a physiological damage threshold. Hence, only disease intensities beyond this threshold yield reduction can be assessed.

Zadoks and Schein (1979), Zadoks (1985) and Nutter et al. (1993a) in descriptive terms classify the various types of losses that farmers, the community and a national economy may sustain. Experimental and statistical methods to assess crop loss are specifically treated in the volume edited by Teng (1987), by Zadoks and Schein (1979) and Oerke et al. (1994). Special aspects of loss assessment are treated in papers by Waggoner and Berger (1987), Hamelink et al. (1988), Campbell and Madden (1990), Schuld (1996) and Bergamin Filho et al. (1997), essentially corroborated Waggoner and Berger's findings. Comparative loss assessment could be done within three categories of loss related to disease intensity and progress: loss magnitudes, loss equivalents and loss profiles. Disease intensity/yield loss relationships are obtained from the assessment of disease intensities (p. 73), remaining healthy plant parts (e.g. leaf area), AUDPC, etc. and appropriate parameters of yield, both sampled from the same plants, plots or fields. Disease severity in percent as a relative term may have to be used with caution in the comparison of data from experiments done under different conditions, as was demonstrated by Rotem et al. (1983) and Waggoner and Berger (1987).

As pointed out in Section 4.3.2.3, disease assessment is subject to errors (Hau et al. 1989) and thus leads to certain degrees of uncertainty in the assessment of yield loss and consequently also in decision-making (see Binns et al. 2000).

Loss magnitudes relate disease intensities to loss in yield measured at harvest as the difference between potential yield (i.e. what a crop could produce when not affected by pests or other constraints), and the actual, harvested yield. However, a disease must not invariably cause damage to a plant (p. 164), e.g. when old affected leaves drop, which are no longer a source to the sink, or when symptoms are mere blemishes which do not affect the market value. Magnitudes may be expressed either in absolute (e.g. weight) or relative (e.g. percent of expected yield) terms for a defined unit loss of yield, monetary values, arable land lost etc. It describes the economic importance of a disease to justify control strategies, research funding etc. and may be of interest to the market and politics. Most of the information on crop loss published so far is on loss magnitudes. Oerke et al. (1994) elaborate on this aspect in detail. Across-studies are needed to consolidate more generally applicable functions for a given pathosystem from the many piecemeal estimates of loss. Quantitative reviews and meta-analysis (Sects. 3.2.2.2 and 3.2.2.3) could be employed to achieve this. Kirby and Archer (1927) did so for stem rust in the USA (see Fig. 7.1) and Kelber (1977) for *Cercospora* leaf spot of sugar beet (Table 3.10). An attempt has been made to quantify yield losses caused by rice pests sensu lato in a range of production situations in tropical Asian countries (Pinnschmidt et al. 1994, 1995a; Savary et al. 2000a,b).

In the comparison of disease intensity/yield loss relationships one could distinguish two types of diseases: (1) those which cause slight losses practically every season with no economic justification for control and (2) diseases which occasionally, often or every year provoke losses beyond the economic damage threshold. The first group of diseases may be called "permanently low loss diseases" and the second one "risk-loss diseases". This difference should be reflected in the loss profile (below). However, comparisons over the years could reveal substantial total losses also caused by the permanently low loss diseases, which may even be higher than a risk-loss one on the same crop.

Loss equivalents relate a unit of disease intensity to a unit of loss in yield (e.g. both in percent; Fig. 7.1). Loss equivalents, like the damage coefficient d in Norton's equation (p. 164), are essential parameters for economic damage thresholds (EDT). These loss equivalents may be assessed, for instance, from linear regressions used to establish a yield/loss relationship. Loss equivalents are also basic to forecasts and risk assessment. Loss equivalents are also essential parameters to design integrated pest management (IPM) schemes that require criteria to make ad hoc decisions for treatments, i.e. an economic (or environmental) damage threshold (EDT). Loss equivalents may be related to those growth stages of the host plant that are critical for yield formation.

Loss profiles (Pinstrup-Andersen et al. 1976, Fig. 7.2) present the proportional effect of the individual constraints in a crop on yield (diseases, pests, weeds, drought, nutrition deficiencies, etc.) and their contributions to yield loss. The con-

Rust severity (%)　　　　　　　　　**Yield Loss (%)**

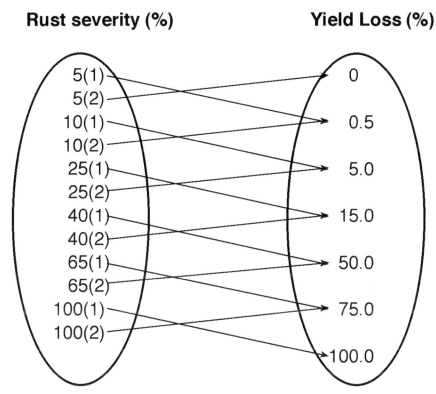

Fig. 7.1. The equivalents of loss (%) incited by disease severity (%) of wheat stem rust (*Puccinia graminis*) in two growth stages of the crop. It is obvious that the same severity may have different loss equivalents. (Data from Kirby and Archer 1927)

straints may be regarded as a subsystem that affects the subsystem host and of the total loss sustained in a crop. Loss profiles ensure a less biased, but more realistic identification of key pests. They are also a better scientific base for the planning or design of effective and durable new, or integrated, pest management systems.

For the loss profile in Fig. 7.2, Pinstrup-Andersen et al. (1976) measured intensities of any obvious constraint in their field experiments on beans and calculated and compared the contribution of each constraint by multivariate regression analysis. The impact of a particular limiting factor on yield from the plots was then estimated by multiplying its regression coefficient by its average market value and then expressing them as proportional loss in kg in Fig. 7.2. There is room for other procedures to be developed, e.g. path coefficient analysis (Fig. 4.14). Another example of the share that each of four wheat disease have in a loss profile with data from a panel of 31 farms which represented 58% of the cultivated area of Idaho (USA) is presented in Fig. 7.3.

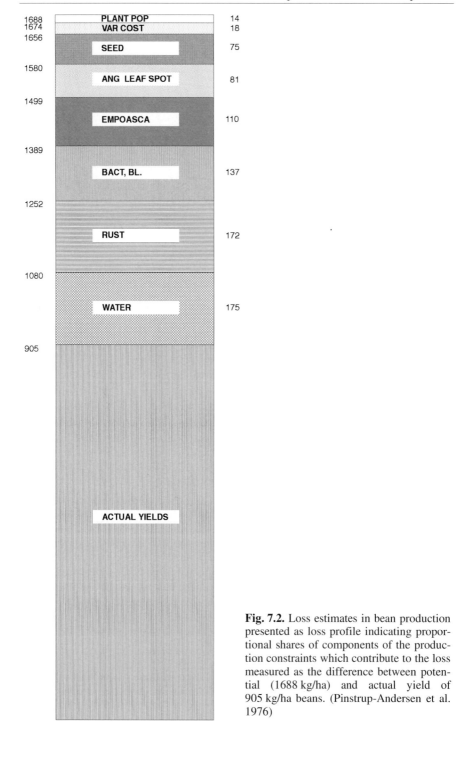

Fig. 7.2. Loss estimates in bean production presented as loss profile indicating proportional shares of components of the production constraints which contribute to the loss measured as the difference between potential (1688 kg/ha) and actual yield of 905 kg/ha beans. (Pinstrup-Andersen et al. 1976)

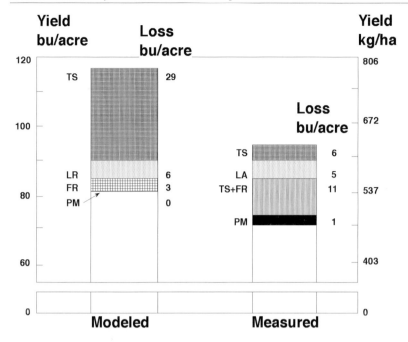

Fig. 7.3. Loss profiles of additive (*measured*) and interactive (*modeled*) representation of yield losses in winter wheat by a complex of four diseases (*TS* tan spot, *LR* leaf rust, *FR* foot rot and *PM* powdery mildew. (Wiese et al. 1984)

Johnson et al. (1986) investigated interactions among constraints for *Alternaria solani*, *Verticillium dahliae*, and the leafhopper *Empoasca fabae* on yields of three potato cultivars. Disease intensity, after artificial infestation, was regulated by applications of the fungicide chlorothalonil and by differential application of insecticides for the natural population of the leafhopper. The somewhat deviating results of the trials over 2 years were due to different levels of overall pest intensity. Maximum yield loss in untreated plots by any combination was 63% in 1983. The maximum yield averaged across the three cultivars by each pest alone was 54 for leafhopper, 31 for early blight and 12% for *Verticillium* wilt. The values for 1984 were 18 for early blight and 15% for leafhopper, when evaluated separately, but combined they only caused 26% yield loss. The effect of rust (*Puccinia arachidis*) and late leaf spot (*Cercosporidium personatum*) on the yield of groundnuts was studied over 2 years by Savary and Zadoks (1992). Chlorothalonil was applied to establish different disease intensities. One of the conclusions reached was that the injury-damage relation differed for both diseases. The late leaf spot affected yield by reducing green leaf area and defoliation, whereas rust acted through "different mechanisms" in addition to the reduction of green leaf area. Both studies used pesticides to establish differential pest intensities, which might have influenced the interactions in one way or another. Untreated fields, like Pinstrup-Andersen et al. (1976) used for syn-epidemiological studies (Table 3.5), would result in less biased estimates of yield loss.

The examples given here on the interaction among constraints and hosts used disease intensities as the measurement term and related them to yield. Perhaps more promising experiments on crop loss relationships could be achieved from estimates of the remaining foliage on diseased plants instead of disease intensities (Hau et al. 1980; Rotem et al. 1983; Waggoner and Berger 1987). Waggoner and Berger (1987) made it clear that important host parameters to estimate yield (rather than yield loss) were the leaf area duration LAD, healthy leaf area duration HLAD, the effective leaf area duration ELAD and the photosynthetically active radiation absorbed by the healthy foliage HAA. Bergamin Filho et al. (1997) proposed that a reference plot that is well protected by fungicides be established to determine the maximum yield for that site. Certainly, the healthy leaf duration (HAD) may be more telling for the assessment of crop losses (Waggoner 1986) for diseases that interfere with the accumulation of assimilates. A pertinent recent example is the three bean diseases for which Schuld (1996) did not find a significant relationship between parameters of disease intensity and yield. Significant relationships, however, existed between the healthy leaf area (HLA) and its integrated HLAD for healthy leaf area duration. Bastiaans (1991) proposed the term "virtual lesion" for the visible lesion and the surrounding tissue already affected by the pathogen. In his field experiments, properly calibrated radiometric measurements gave accurate estimates of LAI when normalised differences (ND) from the reflections of tested wavelengths have been established. Schuld (1996), following Bastiaans (1991) approach, then calculated $LAI = b \; e^{ND}$ where ND is (R810–R610)(R810+R610), for the wavelengths R810 nm and 610 nm. A parameter β described the relation between a virtual and visual lesion. When used in the equation $VBS = 100[1-(1-BS/100)^{\beta}]$, where VBS is the loss incurred by the disease severity BS, already with $\beta > 1$ the virtual disease severity reached 100% before the visual disease severity. The parameter β can be estimated by (1) measurement of gas exchange; (2) radiometry in the field; or (3) from the correlation between HLAD and yield of individual plants (Schuld 1996). The AUDPC, often recommended as a parameter for crop loss assessment, could be employed if its values refer to the growth stages of the host effective to yield formation only.

7.2 Criteria to Compare the Economics of Control Measures: Cost/Return, Thresholds, and Risks

The prevention of a loss in yield calls for control measures. The efficiency of control measures commonly is compared to the increase in yield, and economically by the measurement terms of net return (NR), or gross margin (GM). The cost of control and related measures are weighted against the yield obtained. These costs are important parameters for the calculation of thresholds. An increase in yield alone, possibly obtained at uneconomic costs, cannot be a valid criterion. The NR and GM are defined as (Waibel, pers. comm.)

$NR = [(y \times p) \times (d \times k)] - [(y \times hc) \times (d \times k)] - C$

$$GM=(y{\times}p){\times}\{1-[d(1-k)]\}-(y{\times}hc){\times}\{1-[d(1-k)]\}-C-VC$$

where

y	potential yield, i.e. yield with zero loss per unit area;
p	the expected monetary value per unit yield;
d	yield loss in percent with no control;
k	proportional reduction of yield loss by the control measure;
hc	harvest cost as monetary value per crop unit;
C	costs of control measure or activity;
VC	variable production costs in monetary value per unit area crop.

The NR and GM may be employed as criteria to compare the efficiency of control measures for a season or as a mean over several seasons or years. There are seasons in which measures are not needed to control diseases. If control measures were applied, money was spent unnecessarily and the environment polluted. Hence, more criteria are needed for a valid evaluation and comparison of a standard (or customary) control strategy. These are (1) a threshold concept whether to treat or not and (2) a concept of the risk that such a threshold will be exceeded with diseases that occur under prevailing environment and cultivation practices.

The effectiveness of control programmes or assisting procedures, like forecasts etc., may also be compared by economic criteria. For instance, Phipps et al. (1997) evaluated the benefit of groundnut advisory programmes for growers in Virginia (USA). The programmes covered diseases caused by *Cercospora arachidicola*, *Cercosporidium personatum*, *Sclerotinia minor* and *Calonectria ilicicola* (anamorph *Cylindrocarpon parasiticum*). The programmes consist of an early leaf spot advisory, a Sclerotinia blight advisory, frost advisory, heat units and weather summaries. The latter two advisories are based on weather monitoring networks. The information delivery is achieved by publications, telephone recordings, radio and television broadcasts and an electronic bulletin board. For the two leaf spot diseases, the costs of an established forecast were included in the comparison together with a 5-day Daily Risk Index (DRI) for *Sclerotinia* blight. The parameters of relative humidity and soil temperature formed an environmental index and vine growth and canopy density formed a host index. Both indexes multiplied gave the daily risk index that was then added for five consecutive days to make the 5-day index. The performance of three practices (advisory, 14-day spray schedule or demand, check) was compared with the following criteria for the pathosystems leaf spot and stem diseases:
- number of fungicide applications (N);
- costs ($/ha);
- disease at harvest (% severity for the leaf spots, % incidence for the blight);
- yield (kg/ha);
- gross value ($/ha);
- net return ($/ha), i.e. gross margin value minus total production cost. (see above equations for net return and gross margin, NR and GM).

It would be of interest to compare *thresholds* across pathosystems at the same or different localities, production systems etc. For thresholds, a number of definitions are proposed (Zadoks 1985; Nutter et al. 1993a). We prefer Sylvén's (1968) terminology as criteria for comparative epidemiology, i.e. the economic damage threshold (EDT) and the derived control threshold (CT), either as descriptive or as measurement terms, once pertinent information is available. Any threshold concept, however, will be based on the disease intensity-yield loss relationship. Norton (1976) proposed a rather basic definition of the EDT that has the advantage of clarity as it comprises all relevant parameters for the estimation of this threshold. We amended the equation by a physiological damage threshold PDT. The CTs for plant diseases usually are reached earlier than the EDT (Fig. 7.4). Norton (1976) defines the EDT as

pdik $>$C

and the function to calculate EDT as

EDT $=$C/pdk

and amended here with the physiological threshold PDT

EDT $=$C/pdk+PDT

where

C   cost of control per season and unit area;

EDT   disease intensity (e.g. in percent) as a measurement term for the economic damage threshold;

p   the farm gate price of the harvested crop per unit area and unit yield in monetary units (e.g. $/t/ha);

d   damage coefficient (e.g. t/ha lost per percent disease intensity). The term d could be the regression coefficient b/100 if a linear regression was used to establish a crop loss relationship;

i   the estimated disease intensity;

k   kill coefficient for the reduction of disease intensity and its related loss by the control;

PDT   the disease intensity up to which no reduction in yield can be detected and measured. If the disease intensity/loss relationship is assessed by a linear regression function, then d can be estimated by b/100, and PDT by |a|/b. A negative a value means that the regression line cuts the x-axis at a disease intensity up to which no loss can be measured. PDT practically is a measurement term for what is called "tolerance" of cultivars to disease intensities.

For practical decision-making in the field, the control threshold (Fig. 7.4) is more relevant. The control threshold CT is defined as disease intensity at which a control measure can avoid the economic damage threshold EDT, which is reached after the next incubation period. Hence, CT is the disease intensity which is monitored and the date EDT is (likely) minus the incubation period in days. Figure 7.4 suggests two levels of CT. One, i.e. CT2, can be established from the calculated curve for EDT (e.g. following the amended Norton function), when a known mean disease progress cuts EDT at a certain stage of the epidemic. This control threshold applies when immediate control measures can be taken. However, if there is a time lapse between the monitoring and control due to warning procedures, and preparation and execution of control measures, then a lower disease intensity would be the appropriate control threshold, i.e. CT1, as shown in Fig. 7.4. CT1 would thus be assessed some days earlier (as known from experience) from the mean disease progress curve. The start of the "last" incubation

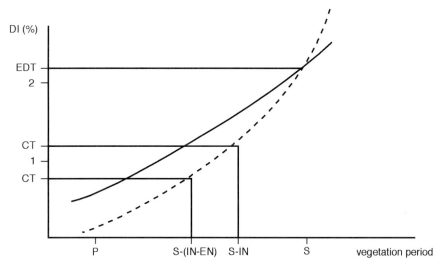

Fig. 7.4. Schematic derivation of control thresholds (CT1 below and CT2 above) for a pro-
tective fungicide from an established EDT curve: a disease progress curve (*broken line*)
known to be common for the pathosystem cuts the EDT curve (*solid line*), for instance, in
growth stage S of the host. The incubation period IN may be 1 week. The disease intensity
1 week before stage S leads to CT2 read from the disease progress curve. After an imple-
mentation phase EN of, say 3 days, CT1 is obtained. (Kranz 1996)

period may be known from forecasts of conditions leading to the next infection
period.

Plant diseases are hazards, a source of danger to crops and their performance,
e.g. yield. The chance or probability that they occur is a risk. For the risk assessment, a calculated, observed or assumed EDT can be a core parameter because it
defines the risk. The other parameter is the "probability" that this event happens.
Probability P for one event is $P=n/N$ with n for the number of risk events within a
total of N events, e.g. n=7 years with disease intensity $\geq$EDT out of N=10 years
observed, P then is 0.7 out of (0. 1). This P applies for a disease A which is permanently present in an area, and $P(A)=1$. If, however, A does not occur every season, then the probability of its occurrence is conditional and the probability that its
disease intensity is $\leq P(B)$. In this case P for one disease then becomes
$P(C)=P(B|A)\times P(A)$; this means that in such a conditional situation the probability
of EDT being reached or surpassed is $P(B)$ provided that A occurs at a probability
$P(A)$. For more than one disease j (or pest j), then $P(C)=\Sigma P(B|A_j)\times P(A_j)$, provided each of them acts independently. When all or some of the A_j sum up to $P=1$,
then a treatment strategy will be justified.

The probabilities P and $P(C)$ are needed for a comparative evaluation of crop
protection strategies. In Table 7.1, two strategies of control, s_1 and s_2, are compared with assumed net return for one disease. With a $P=0.7$, the strategy s_1
"spray" is the better one. If, however, $P=0.3$, then neither strategies differ economically (1140 and 1120 MU respectively) and the choice of the other strategy
may not only be a matter of utility alone, but may be determined by general farm-

Table 7.1. The use of probabilities P to compare the relative advantage of two control strategies in terms of monetary units (MU)[a]. (Adapted from Ogawa et al. 1967)

Event	P	Mean net return (MU) with strategy	
Disease $\geq$EDT		Spray s_1	No spray s_2
E_1 Severity $\geq$EDT	0.7	1000×0.7=700	700×0.7=490
E_2 Severity <EDT	0.3	1200×0.3=360	1300×0.3=390
Mean net return		1060	880

[a]The mean of net return is assumed for a period of 10 years. For E_1 and s_1 a loss of 200 MU represents the residual loss inherent in strategies based on EDT. For E_2 and s_1 the net return is higher because of lower losses, and for E_2 and s_2 100 MU are for the cost of control saved.

ing policy and national politics. The acceptance of risks, and this can be very important, finally depends very much on the individual risk aversion among decision-makers. This psychological aspect may overrule rational considerations, e.g. of utility: a gambler tends to adopt something new or more risky more readily than a timid person. This can make comparisons of what a risk is rather difficult.

More criteria for risk analysis are used when the impact of exotic pests is assessed in plant quarantine, e.g. adaptability of a pathogen to new conditions and its probable rate of spread, expected yield losses and the probability of reaching the EDT, ease and costs of control, and economic, social and cultural aspects for the community or nation.

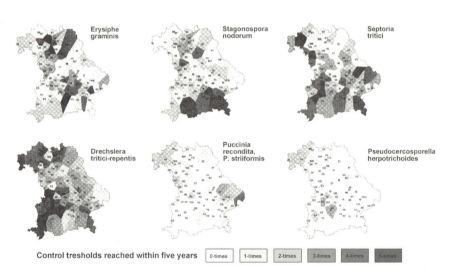

Control tresholds reached in winterwheat during the five years from 1996-2000 in Bavaria

Erysiphe graminis

Stagonospora nodorum

Septoria tritici

Drechslera tritici-repentis

Puccinia recondita, P. striiformis

Pseudocercosporella herpotrichoides

Control tresholds reached within five years   0-times 1-times 2-times 3-times 4-times 5-times

Fig. 7.5. The frequency (0–5 years) of six diseases having reached a defined control threshold in regions of Bavaria during a period of 5 years (1996–2000). The figures in the graphs stand for the number of fields evaluated. In the *white regions* the risk of a disease being controlled by fungicides was nil, the regions marked darkest had to be treated every year during this period. (Tischner and Bauer 2000)

Thresholds, like the one above by Norton, or defined in a different way, can also be used in comparative epidemiology as criteria for the definition of areas of disease risks. Such risk areas have been identified for diseases of barley and wheat in Bavaria, Germany, with their established control thresholds. Figure 7.5 presents six diseases evaluated during the period 1996–2000 (Tischner and Bauer 2000). The criteria they used are frequency of reaching the control threshold in the various regions of the state. The numbers in the graphs for each disease refer to the number of fields observed by the pathogens in relevant growth stages of wheat. Diseases caused by *Erysiphe graminis, Septoria tritici* and *Drechslera tritici-repentis* on wheat within the 5 years most frequently reached the control thresholds in Bavaria.

Table 7.2. Major control measures and their epidemiological effects on the primary disease intensity y_0 and the rate of disease progress r. (After Zadoks and Schein 1979)

Control measure	Major effect on	
	y_0	r
Cultural control		
Crop rotation	+	
Choice of planting date	+	+
Healthy planting material	+	
Eradication of diseased plants and parts	+	+
Elimination of alternate and alternative hosts	+	
Sanitation, surgery	+	
Solarisation	+	
Density of plant stand		+
Soil preparation	+	
Modified plant nutrition		+
Overhead irrigation		+
Physical measures		
Soil treatments, solarisation	+	
Wind shelter, shade etc.		+
Chemical measures		
Seed treatment	+	
Fungicides during vegetation	+	+
Chemical vector control	+	+
Biological measures		
Vertical resistance (VR)	+	
Partial VR	+	+
Horizontal resistance (HR)		+
Cultivar mixtures	+	+
Induced resistance		+
Mycoparasites	+	
Hyperparasites		+
Premunition, putative strains	+	+
Legislative measures		
Quarantine	+	

7.3 Criteria to Compare Efficiency of Disease Control

The notion of "human interference" in the disease square (p. 6) covers intentional control measures as well as any other agricultural practice or policy that may have an unintentional effect on plant diseases and their epidemics.

From the epidemiological point of view, any control action taken affects either the initial or primary disease intensity (y_0), or the rate of disease progress (r), or both, (Table 7.2). Comparisons across control tactics and strategies, or approaches to systems control and design, could, by means of posterior analysis, like meta-analysis (Sect. 3.2.2.3), or equivalence tests (Sect. 3.3.2.2) be appropriate. Results obtained from such comparisons could help in the development of more effective control tactics and strategies. Crop protection measures, like sanitation or the use of cultivars with vertical resistance that reduce y_0, usually delay the start of an epidemic. The effect of a reduced y_0 can be quantified with the sanitation ratio (Van der Plank 1963). The ratio y_0/y_{OS} stands for initial disease without (y_0) and with sanitation (y_{OS}). The delay (Δt) theoretically holds true throughout an epidemic. However, it may change after the first disease cycle as in Table 5.4. Even then, a smaller y_0 in most cases will lead to a lower y_{max}, depending on the mean rate of disease progress during the epidemic. The expected delay (Δt) in days can be calculated with a known or estimated rate r (Van der Plank 1963) for y<0.05 as

$\Delta t = \ln (y_0/y_{OS})/r$, or, if $y_0 \geq 0.05$ (Plaut and Berger 1981), as $\Delta t = [\text{logit} (y_0) - \text{logit} (y_{OS})]/r$

7.3.1 Comparison of Cultural Practices for Disease Control

Specific experimental comparisons on control by cultural practices are apparently rare. Some results have been published which give some clues to their effect; these have been reviewed and compared by Palti (1981; Table 4.10). Wiese (1982) provides some examples for crop management in general. The effects of cultural control on virus diseases (healthy planting material, isolation, windbreaks, barriers and cover crops, mixed cropping, eradication measures) were compared in reviews by Thresh (1982, 1983). The effects of nitrogen or ammonium fertiliser on the intensity of disease caused by a number of pathogens were compared in a comprehensive review by Huber and Watson (1974). The influence of irrigation in its various forms on disease was discussed by Rotem and Palti (1969; Table 4.20). For a given epidemic competence of a pathogen, the effectiveness and potential of cultural control are directly related to the opportunity to manipulate conditions for plant growth (Rotem and Palti 1980). Even with good prospects for disease control, the use of cultural practices often is conditioned by traditions as well as economic, psychological and professional considerations. Also, the relative costs, reliability and speed of alternative control measures to achieve control play a role. Rotem and Palti (1980) by deduction (Sect. 3.3.1) compared several aspects of cultural practices in relation to chemical control (Table 7.3). They also compared economic and human, pathogen and crop, and environmental factors that affect the prospects for cultural practices (Rotem and Palti 1980).

Table 7.3. Comparison of cultural and chemical control measures compared. (Rotem and Palti 1980)

Aspects compared	Cultural control	Chemical control
Results:	Somewhat variable, gradual and unspectacular	More constant and predictable, rapid, often spectacular
Applicability:	Local and specific	Much more general
Know-how required:	Considerable	Almost anybody can spray
Investment:	Mostly low or one-time higher investment	High and recurrent expenses
Field experiments:	Complex, long-term, funds hard to find	Simpler, more rapid, backed by industry

The effects of cultural control measures are generally less spectacular than the use of fungicides and not always easily implemented. Van Bruggen (1995), in a quantitative posterior analysis, compared the plant disease severity in high input to two forms of reduced input (integrated pest management and organic farming systems). Van Bruggen used the observed disease severity as the evaluation criterion (Table 7.4). Cultural control seemed to be particularly effective in organic farming. The other two farming practices did not differ in the means of disease ratings.

Ngugi et al. (2000) compared epidemics of sorghum anthracnose and leaf blight (p. 101) in Kenya. The epidemics of northern leaf blight always started earlier than those of the anthracnose, but the y_{max} for leaf blight at crop maturity was lower. Delayed planting dates shortened the time to disease onset and increased the absolute rate of progress. This earlier and faster epidemic resulted in maximum severity at the milky stage of the crop and at crop maturity. Such an earlier and faster epidemic is thought to be due to higher inoculum available when the crop is sown later. Resistant cultivars for both diseases had the longest delays to disease onset with the slowest absolute rate of disease progress and the lowest disease severities 95 days after emergence. The authors stated that delay in planting

Table 7.4. Relative disease severity[a] in organic (ORG), integrated (INT) and conventional (CONV) wheat farms[b] in the Netherlands. (Adapted from Van Bruggen 1995)

Disease	Pathogen	ORG	IPM	CONV
Leaf rust	*Puccinia recondita*	1–2	1–2	1
Stripe rust	*P. striiformis*	1	2	2
Powdery mildew	*Erysiphe graminis*	0–1	2	2
Leaf blotch	*Mycosphaerella graminicola*[c]	2	2–3	2–3
Glume blotch	*Phaeosphaeria nodorum*[d]	1	1	1
Snow mold	*Microdochium nivalis*	1	2	2
Sharp eyespot	*Rhizoctonia cerealis*	1	2	2
Eyespot	*Pseudocercosporella herpotrichoides*	1	1–2	1–3
Brown foot rot	Fusarium spp.	2	2–3	2–3

[a]0=None, 1=low, 2=moderate, 3=severe; means from several years of observations.
[b]Compiled from experimental data of various authors.
[c]Anamorph *Septoria tritici*.
[d]Anamorph Stagonospora nodorum

affected disease progress of anthracnose more on the resistant cultivars, whereas, in contrast, the progress of leaf blight was more severe on susceptible cultivars. They relate this to higher inoculum in the area and climatic conditions, e.g. dry weather. To screen for resistance in cultivars to both these diseases, the authors concluded that the test should be sown at least 15 days later than normal to obtain the most intense disease. Competition between *C. sublineolum* and *E. turcicum* could not be shown as the coefficients were not significant and their confidence intervals in most cases included zero. Mixtures of the susceptible sorghum cultivar with either the non-host maize or a resistant sorghum cultivar slowed down both the rate of disease progress and the carrying capacity for the anthracnose and the leaf blight, particularly on the latter disease. Intra-row mixtures were more efficient than inter-row mixtures in reducing the disease development. Apparently, the host unit area is the more effective mechanism for the reduction of disease in the two pathosystems.

Various control measures (sowing date and cultivar resistance) were compared in epidemics of chickpea wilt (*Fusarium oxysporum* f. sp. *ciceris*) with disease intensity and time of onset of the epidemic as evaluation criteria (Navas-Cortés et al. 1998; pp. 36, 102). The results first were analysed by principal component analysis (PCA) followed by cluster analysis. For each of the 3 years of field experiments with 216 microplots, the epidemic development was related mainly to the date of sowing. The highest disease intensity and earliest disease onset were closely and positively correlated. A comparison of the interaction of factors studied across epidemics was made in cluster analyses. A dendrogramme (Fig. 7.6) represents the relative similarities obtained among 54 epidemics of *Fusarium* wilt in chickpeas. In the field experiments from 1987 to 1989, three cultivars were used; Cv1 (highly susceptible), Cv2 (moderately susceptible) and Cv3 (susceptible) in 1 year were exposed to the races Foc-0 and Foc-5. These two races had been added to the upper 15 cm of the fumigated soil in the microplots on three dates: 21 December 1987 (SD1), 3 February 1988 (SD2) and 21 March (SD3). The doses of inoculum mixed into the soil were 25 g (IR1), 50 g (IR2) and 100 g (IR3) for race Foc-0, and 6.25 g (IR1), 12.5 g (IR2) and 25 g (IR3) for race Foc-5.

The epidemics of *Fusarium* wilt on chickpeas that were connected after cluster analysis at short linkage distances from the base in Fig. 7.6 had similar measures of the factors. These lower-order clusters comprised those epidemics that occurred in microplots infested with different inoculum doses sown to the three cultivars. The lowest-order clusters grouped epidemics that differed in the initial inoculum dose, irrespective of sowing date, race or cultivar.

Comparative studies on phytosanitation were conducted within the framework of an international co-operative programme on witches'-broom of cacao (Maddison et al. 1993). The objective was to evaluate the efficacy of broom removal and to measure disease gradients of witches'-broom disease. "In general, the number of symptoms on flower cushions were reduced more readily by sanitation than those on shoots and pods. At a given site, the slope of the curve for cushion symptoms was steeper and the levels of 'background' infection were reduced more, than for shoot and pod infections".

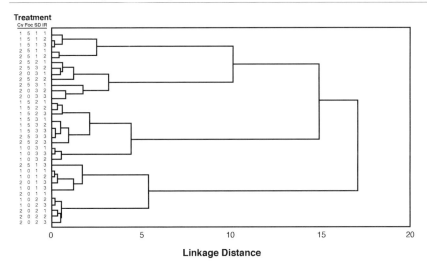

Fig. 7.6. A dendrogramme representing relative similarities among 54 *Fusarium* wilt epidemics (*Fusarium oxysporum* f. sp. *ciceri*) for 1 year with three chickpea cultivars, Cv1, Cv2 and Cv3, sown at three dates (SD1–3) in microplots artificially infected with races Foc 0 and Foc 5 with three inoculum doses (IR 1–3, see text). (Navas-Cortés et al. 1998)

7.3.2 Comparison of Changes of Pathotype Frequencies in Populations of Hosts

A substantial number of comparisons have been published at the micro-level of epidemiology, namely on the distribution and occurrence of virulence and pathotypes (races) of pathogens as well as on their components (Table 4.15). Race patterns, frequencies and diversity coefficients are used as measurement terms (Table 4.17).

Criteria to be used for analyses of changes in pathotypes (see Welz 1988) are aerospora catches by spore samplers. An alternative method is the catching of inoculum on (juvenile) susceptible host plants. These catches can be subsequently cultivated and their race identified later. Sampling in crops is done from lesions collected in the field. During the exodemic, the lesions from organs (e.g. the flag leaves) may be chosen as the sample units to assess the proportion of allosporae that have landed. Later, during the esodemic, samples should be taken from inside the canopy. The identification of the pathotypes may be done with sets of host differentials. An agreed-upon nomenclature of the races should be used that can assign only one virulence genotype to a name, like the binomial Habgood (1970) key. Identification also may be possible by molecular methods. When frequencies of pathotypes in pathogen populations are plotted over the years, changes due to the replacement of cultivars with differential R-genomes (e.g. vertical resistance) can be compared (Fig. 7.7). Despite fluctuations, a clear trend is evident. The pathotype 23 was dominant as long as cultivar Aura (with the R-genes Mlg, Mla6) occupied 60% of the spring barley acreage in the state of Hesse between 1983 and 1986. Thereafter, other pathotypes gradually emerged.

Cultivars with specific resistance genes that are exposed early in the vegetation period to the dispersed and overseasoning inoculum can screen for matching virulence genes. This reduces y_0. These cultivars delay the (visible) onset of the epidemics compared to cultivars with no specific resistance genes. The reduction

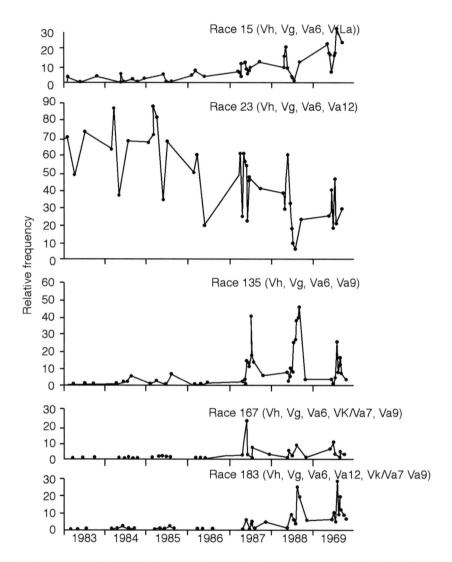

Fig. 7.7. Relative frequencies (%) of five races of *Erysiphe graminis* f. sp. *hordei* in Hesse, Germany, between 1983 and 1991. Compiled from data of various authors (Welz and Kranz 1997). Samples were collected from susceptible spring and winter barley cultivars. With race-specific resistance (VR), cleistothecia, and volunteer seedlings

of early disease intensity is more or less proportional to the frequency of matching virulence in the cultivars with specific resistance genes (see Δt, p. 167). A short generation time (latent period) and a large reproductive value (progeny/parent ratio) may compensate for the delay Δt and a rather rapid buildup of an effective, virulent population of the pathogen can occur. Retarded disease progress in resistant cultivars, e.g. slow rusting, usually leads to a lower y_{max}. Life table studies (Fig. 4.6) were done to identify the component of the infection chain that most reliably determined retardation in the epidemic. The longer duration of the latent period compared to a standard cultivar is used in the early screening processes in resistance breeding.

Multi-lines and cultivar mixtures were introduced as control mechanisms to operate against dominating virulence and the dreaded decline of resistance in cultivars. These strategies combine three principles that are also known from populations of wild host plants. These are: (1) decrease in spatial density of susceptible host plants (probably the most important one). This particularly affects pathogens with shallow gradients like windborne diseases, e.g. *Erysiphe graminis* that are more easily controlled by mixtures. (2) Barrier effects with non-matching host genotypes in which that pathogen can collect and thus alloinoculum is removed. Therefore, the inoculum cannot reach plants with matching genotypes. Finally, (3) induced resistance may be caused by non-pathogenic spores that land on host plants. The decrease in disease intensity in multi-lines and cultivar mixtures is partly due to reduced rates of disease increase (Wolfe 1986; Wolfe and Finckh 1997; Finckh and Wolfe 1998). The efficacy of cultivar mixtures is compared by disease intensities primarily, however, also by the increase in yield of the mixtures (Table 7.5). From the 122 trials analysed in Table 7.5, 50 mixtures with two-component cultivars had a higher yield, and 60 mixtures with three cultivars had higher yields than the single components of the mixtures. The main effect in this comparison was due to the reduction of powdery mildew against which resistance was largely effective.

Table 7.5. Summary of mean yield (t/ha) comparisons from field trials in England over 11 years comparing different mixtures of three barley cultivars with three components grown as pure stands (Wolfe and Finckh 1997)

No. of trials	Mean yield of mixtures	Mean yield of components	Percentage increase in mixtures
122	5.61	5.20	7.88

With an EPIMUL-based simulator, Lannou and Mundt (1996) found that in mixtures the selection for simple or complex races in the pathogen population may not depend on initial race frequencies. For a given multiplication rate, the complex race frequency increased faster when (1) the spore dispersal gradient was shallow; (2) distribution of initial disease was generalised; (3) the amount of initial disease was reduced; and (4) when the number of mixture components was increased. This was attributed to a better efficacy of the mixture to control simple races, which results in a higher relative fitness of the complex race. Xu and Ridout

(2000) undertook a stochastic simulation of the spread of race-specific and race-non-specific aerial fungal pathogens in cultivar mixtures with the variables of sporulation rate, median spore dispersal distance, probability of cross-infection among hosts and pathogen races, the number of host genotypes as mixture components and the spatial arrangement of the mixture components. The disease dynamics were measured by the rate parameter of the logistic equation and the AUDPC for incidence and severity. In their simulations, mixtures were more effective in reducing race-specific pathogens than race-non-specific pathogens. These effects were mainly due to the number of race-specific components or, for race-non-specific mixtures, the proportion of susceptible genotypes in the mixtures and their spatial arrangements. A higher sporulation rate decreased mixture efficacy. For race-specific mixtures, the disease intensity decreased with increasing number of components. The shorter the spore dispersal distance, the less effective the mixture.

With the rate-reducing. non-differential and quantitative resistance QTL (quantitative trait-linked loci), the level of aggressiveness selects the pathogen genotypes capable of overcoming a given level of non-qualitative resistance. QTL is typically a partial, polygenic resistance and can be measured on a continuous quantitative scale (e.g. disease intensity in percent). In ANOVA, significant main effects stand for "horizontal" resistance and the significant interaction of cultivars × isolates for "vertical" resistance (Vanderplank 1984). ANOVA could also be employed to reveal genotype-by-environment interactions that are frequently observed with quantitative resistance, with other quantitative traits across locations or with genetically different pathogen populations (Geiger and Heun 1989).

The reduction of the rate of epidemic increase commonly is related to durable resistance being a desirable objective in plant breeding for use in crop protection (Johnson 1984; Parlevliet 1997). The durability of cultivar resistance is also known with incomplete and complete vertical resistance when certain conditions prevail. Table 7.6 lists possible criteria for comparisons across pathosystems, host genotypes etc..

Table 7.6. Factors that favour the durability of vertical or specific resistance

---

Low epidemic competence of the pathogen[a]

Permanently unfavourable climatic conditions for high disease intensities and rates
Wide crop rotation
Diversity[b] of R-genes cropped
Saprophytic phases which alter pathotype frequencies
Regular and effective use of fungicides

---

[a]See Table 4.14.
[b]See Table 4.17.

7.3.3 Comparison of Chemical Control

7.3.3.1 Efficacy of Fungicides

Fungicides act to reduce the initial disease intensity y_0 or to slow down the rate r (Table 7.3). The efficacy of fungicides can be enhanced by other crop protection measures, e.g. crop rotation, choice of cultivar and nitrogen fertilisation. The interactive effects of treatment combinations may vary somewhat from year to year (Herrman and Wiese 1985) and from site to site.

Major criteria evaluate and compare the efficacy of fungicides as the inoculum destroyed which is commonly measured by means of the ED_{50}, or "kill efficiency" as in the Norton equation for the EDT (p. 164). Another criterion is the duration of effectiveness, or the persistence of the compound (Fig. 3.1).

Seed and soil treatments reduce the overseasoning inoculum Q, inhibit infection of emerging plants (y_0) or render juvenile plants resistant for some weeks. The efficacy of fungicides on rates, according to Van der Plank (1963), depends on iR, the basic infection rate (see also Berger 1988). This factor determines the proportion of spores that must be killed to keep iR<l and thus prevent the start of an epidemic. The lower the iR, then the easier a disease would be controlled by fungicides because the absolute amount of spores killed is smaller (Van der Plank 1963). Thus, late blight with an iR of 7 is more conveniently controlled than wheat stem rust with an iR of 360. Apple scab (*Venturia inaequalis*) has a rather long infectious period i. For this reason, it can be controlled only if R is kept small, e.g. by controlling new infections. Diseases with a tendency to be explosive should be treated as early as possible, even with preventive measures. For slow developing diseases Rotem (pers. comm.) suggests treatment be postponed until the host turns more susceptible. Early blight of potato (*Alternaria solani*) was controlled more effectively by early applications than was late blight (*Phytophthora infestans*), particularly when the latter was severe (Shtienberg and Fry 1990). This experimental result was corroborated by computer simulation.

The fungicidal side effects of Anilazin, Carbendazim and Pyrazophos on non-target *Septoria* spp., *Puccinia recondita* and *P. striiformis* were studied for 2 years on wheat in field experiments. The results were not striking, although most treatments that received the fungicidal spray were statistically significant from the untreated (Weber and Kranz 1994b). The side effects on non-target insect pests were clearer (Weber and Kranz 1994a). Pyrazophos favoured aphids and thrips, but decreased the incidence of *Oulema* spp. and saddle midges. The application of Analazin resulted in a higher incidence of *Contarinia tritici* and *Sitodiplosis mosellana* and inhibited aphids. The broad-spectrum fungicides Propiconazol and Fenpropimorph inhibited *Thysanoptera* and gall midges, but favoured an increase in aphids and *Oulema* spp. The fungicides also affected the composition of pests observed on the leaves which changed among the three cultivars and between years. Fungicides may cause iatrogenic effects (Horsfall 1979). Griffiths (1981) discussed a number of agrochemicals which favoured "new" diseases that otherwise would not have become a problem. This may happen through an affect on (1) the host plant; (2) the pathogen or (3) the ecosystem in which host and pathogen exist.

The control by fungicides of pathogens that infect their hosts in the soil is more difficult and less satisfactory. Aspects of the management of soilborne diseases are compared across two pathosystems by Morton (1994). Matheron et al. (1997) investigated the effect of fungicide treatment on the health of citrus trees in six citrus plantations and three soil depths over 7 years. The relation of the distribution and seasonal population changes of *Phytophthora citrophthora* and *P. parasitica* was of particular interest. This extended treatment programme with fosetyl-Al and metalaxyl resulted in significantly healthier tree canopies and higher root densities compared to non-treated trees. However, no significant and consistent difference was detected in the population densities of either *Phytophthora* species. No significant correlation was found between soil temperature or soil moisture and inoculum density of either pathogen at the time of sampling.

7.3.3.2 Fungicide Resistance

There are detailed accounts on fungicide resistance by Georgopoulos and Dekker (1982). Staub (1991) reviewed extensively the history and factors that affect the development of resistance in major groups of risk fungicides. Dekker (1988) described the methods to measure fungicide resistance. Selection of fungicide-resistant strains in a pathogen population by risk fungicides follows the same principles as discussed in Section 4.4.3) and are applicable to across-studies as well. Fungicide-sensitive and fungicide-resistant pathogen genotypes are selected instead of virulence. Selection processes can either be directional, leading to a gradual selection of fungicide-resistant subpopulations (strains) or disruptive, with the sudden appearance of a resistant population to a new fungicide. Disruptive selection may result from a high proportion of strains in a pathogen population that are resistant to a compound before it comes into use. Or genotypes with higher fitness that survive the recommended dose of the fungicide will build up rapidly in the pathogen population because of their higher rates.

Common parameters to compare the degree of fungicide resistance are the LD_{50}, the proportion of sensitive and resistant strains in the population and their relative fitness, rates of fungicide-resistant versus sensitive strains (subpopulations) and a standard selection time t_s (see below). Usually, the infection rate r is used as a fitness parameter to measure the behaviour of either the total populations or its sensitive and resistant subpopulations. The standard selection time t_s is affected, apart from the specific selection efficacy of a compound, by the way and frequency a risk fungicide is applied to crops. The parameters are: number of spray rounds per season, seasons in succession, area applied to, the quality of spraying (e.g. coverage of host surface), dose and whether applied as a mixture or alternating with other fungicides.

Skylakakis (1983) proposed a set of three differential equations to describe the disease increase in a pathogen population with fungicide-resistant and sensitive subpopulations. These are

$$dx/dt = r_1 x(1-x-y)$$

$$dy/dt = r_2 y(1-x-y)$$

$dz/dt = r_3 z(1-x-y)$

where x and r_1 are disease intensity and the rate for the sensitive subpopulation, y and r_2 are the respective values for the resistant subpopulation and $z = x + y$ and r_3 are the parameters for the total population of the pathogen, hence $dz/dt = dx/dt + dy/dt$ which leads to

$r_3 = (x r_1 + y r_2)/(x+y)$

This equation, though it may be used to compare the impact on disease control, is a measure of instant values "because, even if r_1 and r_2 did not change with time, but were different from each other, r_3 would change as the ratio x/y changes with time" (Skylakakis 1983).

Skylakakis (1983) also proposed a standard selection time $t_s = 1/(r_2 - r_1)$ as the time necessary for the resistant subpopulation to multiply exponentially. Thus, it is a parameter to estimate (and compare) how long it will take resistant strains of the pathogen, starting from a certain initial level, to render a pathogen population fully resistant. This time lapse (Table 7.7) then can be calculated as (Skylakakis 1983)

$\ln(y_t/x_t) = \ln(y_0/x_0) + (r_2 - r_1)t_s.$

Table 7.7. Standard selection times t_s for fit[a], highly resistant[b] populations of selected pathogens in presence of an efficient[c] fungicide. (Skylakakis 1983)

Pathogen	Infection rate (unit/day)	Latent period (days)	Standard selection time t_s (days)
Phytophthora infestans	0.400[d]	5	4.1
Puccinia graminis	0.400	10	6.2
Erysiphe graminis	0.200	8	7.8
Cercospora beticola	0.100	15	4.3
Ustilago spp.	0.008	–	s[-267.3][e]

[a]Selection rate S in the absence of the fungicide is S=0.01.
[b]No effect of fungicide on basic infection rate R of resistant population is assumed.
[c]90% reduction of basic infection rate of sensitive subpopulation caused by fungicide is assumed on all pathogens, except *Ustilago* spp. with a 95% reduction.
[d]Infection rates from various sources.
[e]Negative sign indicates that selection will still operate in favour of the sensitive subpopulation, but slower, with a selection coefficient S=0.004 instead of S=0.01.

Fungicide resistance evolves more slowly, when the following conditions apply as criteria; (1) initial frequency of an already resistant strain is reduced; (2) the rates r for both, resistant and sensitive subpopulations are reduced and (3) the rate for the resistant subpopulation is reduced relative to the sensitive subpopulation (Milgroom and Fry 1988). Little is known about the longevity of genotypes resistant to fungicide, after their proportion in the population has been reduced and whether their fitness relative to the sensitive genotypes is lower. Benzimidazol resistance in *Botrytis cinerea* on grapes in the Alsace was still widespread 10 years after the use of this fungicide had ceased, whereas resistance to dicarboximides was confined to small pockets only (Staub 1991).

When it is assumed that a pathogen population is constantly exposed to a risk fungicide, this equation estimates, with an assumed initial proportion of 10% resistant strains, a standard selection time of 30–80 days for *Phytophthora infestans* with a t_S of 4.1 days. For the slower *Cercospora beticola* with t_S=14.3 days, this would last 14.3/4.1=3.49 times longer. A comparison of standard selection times for some pathogens is shown in Table 7.7.

Pathotypes are subpopulations with virulence to a particular resistance genotype. Within such subpopulations, there are genotypes with various levels of ED_{50} for a risk fungicide that are selected differentially. With continued use of a risk fungicide, genotypes with higher sensitivity (i.e. low ED_{50}) are gradually eliminated. Then, only genotypes with low sensitivity (high ED_{50}) remain in the population of the pathotype or the entire gene pool of the pathogens (Wohlleber et al. 1993; Pons-Kühnemann 1994). This leads to the phenomenon of fungicide resistance of a pathogen population. This sets the stage for a vicious circle (Welz and Kranz 1997). After a resistant cultivar becomes very popular, virulent pathotypes appear and the cultivar appears to be less resistant. A risk fungicide will be applied to control the disease. Subsequently, the risk fungicide also loses effectiveness through elimination of the more sensitive pathogen genotypes. Eventually, the cultivar will be as susceptible as before the use of the fungicide and will be replaced by another resistant cultivar, or another fungicide, whatever is more readily available and economic.

References

Amorim L (1984) Contribution à l'analyse des épidémies de la tavelure du pommier. Stabilité des résistances génétiques de l'hôte et la valeur adaptive des pathotypes. Thèse Doct Ing Univ Paris-Sud Orsay

Amorim L, Bergamin Filho A (1991a) Sugarcane smut development models. I. Annual curves of disease progress. Z Pflanzenkr Pflanzenschutz 98:605–612

Amorim L, Bergamin Filho A (1991b) Sugarcane smut development models. II. Polyetic curves of disease progress. Z Pflanzenkr Pflanzenschutz 98:613–618

Amorim L, Bergamin Filho A, Hau B (1993) Analysis of progress curves of sugarcane smut on different cultivars using functions of double sigmoid curves. Phytopathology 83:933–936

Amorim L, Berger RD, Bergamin Filho A, Hau B, Weber GE, Bacchi LMA, Vale FXR, Silva MB (1995) A simulation model to describe epidemics of rust of Phaseolus beans. II. Validation. Phytopathology 85:722–727

Analytis S (1973) Methodik der Analyse von Epidemien dargestellt am Apfelschorf (*Venturia inaequalis* (Cooke).Aderh.). Acta Phytomed 1:76 pp

Analytis S (1977) Über die Relation zwischen biologischer Entwicklung und Temperatur bei phytopathogenen Pilzen. Phytopathol Z 90:64–76

Analytis S (1979) Die Transformation von Befallswerten in der quantitativen Phytopathologie. II. Phytopathol Z 96:156–171

Analytis S, Kranz J (1972) Über die Korrelation zwischen Befallshäufigkeit und Befallsstärke von Pflanzenkrankheiten. Phytopathol Z 73:201–207

Apel H, Herrmann A, Richter O (1999) Ein Entscheidungshilfesystem für IPM vom *Helicoverpa armigera* in den Tropen und Subtropen auf der Basis eines regelbasierten Fuzzy-Modells. Z Agrarinform 4:83–90

Aust HJ (1981) Über den Verlauf von Mehltauepidemien innerhalb des Agro-Ökosystems Gerstenfeld. Acta Phytomedica 7. Parey, Berlin

Aust HJ, Hau B (1981) Einfluss kompensierender Wirkungen auf die Latenzzeit von *Septoria nodorum*. Z Pflanzenkr Pflanzenschutz 88:655–664

Aust HJ, Hau B (1983) Die Latenzzeit von *Septoria nodorum* in Abhängigkeit von der ontogenetisch bedingten Anfälligkeit des Sommerweizens. Z Pflanzenkr Pflanzenschutz 90:55–62

Aust HJ, Kranz J (1988) Experiments and procedures in epidemiological field studies. In: Kranz J, Rotem J (eds) Experimental techniques in plant disease epidemiology. Springer, Berlin Heidelberg New York, pp 7–17

Aust HJ, Bashi E, Rotem J (1980) Flexibility of plant pathogens in exploiting ecological and biotic conditions in the development of epidemics. In: Palti J, Kranz J (eds) Comparative epidemiology. A tool for better disease management. PUDOC, Wageningen, pp 46–56

Aust HJ, Hau B, Kranz J (1985) Regelung von Schaderregerpopulationen in natürlichen sowie in Agro-Ökosystemen. Ber Dtsch Bot Ges 98:199–208

Aylor DA (1986) A framework for examining inter-regional aerial transport of fungal spores. Agric For Meteorol 38:263–288

Aylor DA, Taylor GS, Raynor GS (1982) Long-range transport of tobacco blue mold spores. Agric Meteorol 27:217–230

Baker KF, Smith SH (1966) Dynamics of seed transmission of plant pathogens. Annu Rev Phytopathol 4:311–334

Bardossy A, Duckstein L (1995) Fuzzy rule-based modeling with applications to geophysical. Biological and engineering systems. CRC Press, Boca Raton

Barker KR, Noe JP (1988) Techniques in quantitative nematology. In: Kranz J, Rotem J (eds) Experimental techniques in plant disease epidemiology. Springer, Berlin Heidelberg New York, pp 223–236

Bashi E, Rotem J (1974) Adaptation of four pathogens to semi-arid habitats as conditioned by penetration rate and germinating spore survival. Phytopathology 64:1035–1039

Bashi E, Rotem J (1975) Host and biotic factors affecting sporulation of *Stemphylium botryosum* f. sp. *lycopersici* on tomatoes and of *Alternaria porri* f. sp. *solani* on potatoes. Phytoparasitica 3:27–38

Bassanezi RB, Amorim L, Bergamin-Filho A, Hau B (1998) Effects of bean line pattern mosaic virus on the monocyclic components of rust and angular leaf spot of Phaseolus bean at different temperatures. Plant Pathol 47:298–298

Bastiaans L (1991) Ratio between virtual and visual lesion size as a measure to describe reduction in leaf photosynthesis of rice due to leaf blast. Phytopathology 81:611–615

Batschelet E (1971) Introduction to mathematics for life scientists. Springer, Berlin Heidelberg New York

Benson DM (1994) Inoculum. In: Campbell CL, Benson DM (eds) Epidemiology and management of root diseases. Springer, Berlin Heidelberg New York

Bergamin Filho A, Amorim L (1996) Doencas de plantas tropicais. Epidemiologia e controle económico. Editoria Agronómica Ceres Lda, Sao Paulo

Bergamin Filho A, Carneiro SMTPG, Godoy CV, Amorim L, Berger R, Hau B (1997) Angular leaf spot Phaseolus beans: relationship between disease, healthy leaf area, and yield. Phytopathology 87:506–515

Bergamin Filho A, Amorim L, Larranjeira FF, Berger RD, Hau B (1998) Análise temporal do amarelecimento fatal do dendezeiro como ferramenta para elucidar sua etiologia. Fitopatol Brasil 23:391–396

Berger RD (1973) Early blight of celery: analysis of disease spread in Florida. Phytopathology 63:1161–1165

Berger RD (1974) A guide to spraying sweet corn for *Helminthosporium* blight by likely disease spread. Univ Fl Agric Res Educ Cent, Belle Glade Res Rept EV-1974-I5, 3 pp

Berger RD (1975) Disease incidence and infection rates of *Cercospora apii* in plant spacing trials. Phytopathology 65:485–487

Berger RD (1981) Comparison of the Gompertz and logistic equations to describe plant disease progress. Phytopathology 71:716–719

Berger RD (1988. The analysis of effects of control measures on the development of epidemics. In: Kranz J, Rotem J (eds) Experimental techniques in plant disease epidemiology. Springer, Berlin Heidelberg New York, pp 137–151

Berger RD, Bartz JA (1982) Analysis of monocyclic pathosystems with *Erwinia-Lycopersicon* as model. Phytopathology 72:365–369

Berger RD, Luke HH (1979) Spatial and temporal spread of oat crown rust. Phytopathology 69:1199–1201

Berger RD, Hau B, Weber GE, Bacchi LMA, Bergamin Filho A, Amorim L (1995) A simulation model to describe epidemics of rust of Phaseolus beans. I. Development of the model and sensitivity analysis. Phytopathology 85:715–721

Berger RD, Bergamin Filho A, Amorim L (1997) Lesion expansion as an epidemic component. Phytopathology 87:1005–1013

Binns MR, Nyrop JP, van der Werf W (2000) Sampling and monitoring in crop protection. The theoretical basis for developing practical decision guides. CAB Publishing, CAB International Wallingford, UK

Blettner M, Sauerbrei W, Schlehofer B, Scheuchenpflug T, Friedenreich C (1997) Vergleich von traditionellen Reviews, Metanalysen und gepoolten Analysen zur Bewertung von Risikofaktoren. Inform Biometr Epidemiol Med Biol 28:148–166

Bock KR (1962) Seasonal periodicity of coffee leaf rust and factors affecting the severity of outbreaks in Kenya Colony. Trans Br Mycol Soc 45:289–300

Börner H (1990) Pflanzenkrankheiten und Pflanzenschutz, 6th edn. UTB 518. Ulmer, Stuttgart

Bonfig-Picard G (1982) Epidemiologische Feldstudien über Wechselwirkungen von Schadorganismen im Agro-Ökosystem Weizen. PhD Thesis, Univ Giessen

Bonfig-Picard G, Kranz J (1984) Untersuchungen über Wechselwirkungen von Schadorganismen im Agro-Ökosystem Weizen. Z Pflanzenkr Pflanzenschutz 91:619–628

Burdon JJ (1993) The structure of pathogen populations in natural plant communities. Annu Rev Phytopathol 31:305–325

Burdon JJ, Chilvers GA (1982) Host density as a factor in plant disease ecology. Annu Rev Phytopathol 20:143–166

Butt DJ, Royle DJ (1980) The importance of terms and definitions for a conceptually unified epidemiology. In: Palti J, Kranz J (eds) Comparative epidemiology. A tool for better disease management. PUDOC, Wageningen, pp 29–45

Camann MA, Culbreath AK, Pickering J, Todd JW, Demski JW (1995) Spatial and temporal patterns of spotted wilt epidemics in peanut. Phytopathology 85:879–885

Campbell CL (1986) Interpretation and uses of disease progress curves for root diseases. In: Leonard KJ, Fry WE (eds) Plant disease epidemiology. Population dynamics and management, vol 1. Macmillan, New York, pp 38–54

Campbell CL (1998) Disease progress in time: modelling and data analysis. In: Garreth-Jones D (ed) The epidemiology of plant diseases. Kluwer, Dordrecht, pp 181–206

Campbell CL, Benson DM (1994a) Epidemiology and management of root diseases. In: Campbell CL, Benson DM (eds) Springer, Berlin Heidelberg New York

Campbell CL, Benson DM (1994b) Spatial aspects of the development of root disease epidemics. In: Campbell CL, Benson DM (eds) Epidemiology and management of root diseases. Springer, Berlin Heidelberg New York, pp 195–243

Campbell CL, Madden LV (1990) Introduction to plant disease epidemiology. Wiley, New York

Campbell CL, Madden LV, Pennypacker SP (1980) Structural characterization of bean root rot epidemics. Phytopathology 70:152–155

Campbell CL, Jacobi WR, Powell NT, Main CE (1984) Analysis of disease progression and the randomness of occurrence of infected plants during tobacco black shank epidemics. Phytopathology 74:230–235

Carter W (1973) Insects in relation to plant disease, 2nd edn. Wiley, New York

Cheverud JM, Dow MM, Leutenegger W (1985) The qualitative assessment of phylogenetic constraints in comparative analyses: sexual dimorphism in body weight among primates. Evolution 39:1335–1351

Chiarappa L (1881) Crop loss assessment methods. Supplement 3. FAO/CAB

Cliff AD, Ord JK (1981) Spatial processes: models and application. Pion, London

Clifford HT, Williams WT (1976) Similarity measures. In: Williams WT (ed) Pattern analysis in agricultural sciences. Elsevier, Amsterdam

Coakley SM, Line RF, McDaniel LR (1988) Predicting stripe rust severity on winter wheat using an improved method for analyzing meteorological and rust data. Phytopathology 78:543–550

Cohen Y, Rotem J (1971) Field and growth chamber approach to epidemiology of *Pseudoperonospora cubensis* on cucumber. Phytopathology 61:736–737

Cooke BM, Jones DG (1970) The epidemiology of *Septoria tritici* and *S. nodorum* II. Comparative studies of head infection by *Septoria tritici* and *S. nodorum* on spring wheat. Trans Br Mycol Soc 54:395–404

Cox AE, Large EC (1960) Potato blight epidemics throughout the world. US Dept Agric Handbook 174

Cox DR, Isham V, Northrop P (2000) Statistical modeling and analysis of spatial patterns. In: Dieckmann U, Law R, Metz JAJ (eds) The geometry of ecological interactions. Simplifying spatial complexity. Cambridge Univ Press, Cambridge, pp 65–88

Cracraft J (1981) The use of functional and adaptive criteria in phylogenic systematics. Am Zool 21:21–36

Croxall HE, Smith LP (1976) The epidemiology of potato blight in the East Midlands. Ann Appl Biol 82:451–466

Culbreath AK, Beute MK, Campbell CL (1991) Spatial and temporal aspects of *Cylindrocladium* black rot in resistant and susceptible peanut genotypes. Phytopathology 81:144–150

Davis JM (1987) Modeling the long-range transport of plant pathogens in the atmosphere. Annu Rev Phytopathol 25:169–188

Davis JM, Main CE (1986) Applying atmospheric trajectories analysis to problems in epidemiology. Plant Dis 70:490–497

Dekker J (1988) How to detect and measure fungicide resistance. In: Kranz J, Rotem J (eds) Experimental techniques in plant disease epidemiology. Springer, Berlin Heidelberg New York, pp 137–151

De Vallavieille-Pope C, Huber L, Leconte M, Goyeau H (1995) Comparative effects of temperature and interrupted wet periods on germination, penetration, and infection of *Puccinia recondita* f.sp. *tritici* and *P. striiformis* urediospores on wheat seedlings. Phytopathology 85:409–415

Dickersin K, Berlin JA (1992) Meta-analysis: state-of-the-science. Epidemiol Rev 14:154–176

Dickinson CH, Lucas JA (1977) Plant pathology and plant pathogens. Blackwell, Oxford

Dieckmann UR, Law R, Metz JAJ (2000) The geometry of ecological interactions. Simplifying spatial complexity. Cambridge Univ Press, Cambridge

Dinoor A, Eshed N (1984) The role and importance of pathogens in natural plant communities. Annu Rev Phytopathol 22:443–466

Disthaporn S, Hau B, Kranz J (1993) Comparison of sampling procedures for rice diseases, blast and tungro. Plant Pathol 42:313–323

Duck NB, Bonde MR, Petersen GL, Bean GA (1987) Sporulation of *Perenosclerospora sorghi*, *P. sacchari*, and *P. philippensis* on maize. Phytopathology 77:438–441

Dutzmann S (1985) Zur Analyse der Beziehungen zwischen Klimadaten und Sporenproduktion und Sporenverbreitung von *Erysiphe graminis* f. sp. *hordei*. Z Pflanzenkr Pflanzenschutz 92:629–642

Elze M, Blume HH (1999) Bioequivalence trials – status and perspectives. Inform Biometr Epidemiol Med Biol 30:87–95

Erickson WP, McDonald LL (1995) Tests of bioequivalence of control media and test media in studies of toxicity. Environ Toxicol Chem 14:1247–1256

Fargette D, Muniyappa V, Fauquet C, Nguessan P, Thouvenel JC (1993) Comparative epidemiology of 3 tropical whitefly-transmitted geminiviruses. Biochimie 75:547–554

Fatemi F, Fitt BDL (1983) Dispersal of *Pseudocercosporella herpotrichoides* on *Pseudopeziza brassicae* spores in splash droplets. Plant Pathol 32:401–404

Ferrandino FJ (1993) Dispersive epidemic waves. I. Focus expansion within a linear planting. Phytopathology 83:795–802

Ferrandino FJ (1996) Two-dimensional distance class analysis of disease incidence data: problems and possible solutions. Phytopathology 86:685–691

Ferrandino FJ (1998) Past nonrandomness and aggregation to spatial correlation: 2DCORR, a new approach for discrete data. Phytopathology 88:84–91

Finckh MR, Wolfe MS (1998) Diversification strategies. In: Garreth-Jones D (ed) The epidemiology of plant diseases. Kluwer, Dordrecht, pp 231–259

Fitt BD, McCartney L (1986) Spore dispersal in relation to epidemic models. In: Leonard KJ, Fry WE (eds) Plant disease epidemiology. Population dynamics and management, vol 1. Macmillan, New York, pp 311–345

Fitt BD, Gregory PH, Todd AD, McCartney HA, MacDonald OC (1987) Spore dispersal and plant disease gradients: a comparison between two empirical models. Phytopathol Z 118:227–242

Fleiss JL (1993) The statistical basis of meta-analysis. Stat Methods Med Res 2:121–145

Fouré E (1982) Les cercosperioses du bananier et leurs traitement. Comportement de variétés. Etude de la sensibilité variétale des bananiers et plantains. *Mycosphaerella fijiensis* au Gabon. Fruits 37:749–766

Francl LJ, Neher DA (eds) (1997) Exercises in plant disease epidemiology. APS Press, St Paul, Minnesota

Fravel DR, Engelkes CA (1994) Biological management. In: Campbell CL, Benson DM (eds) Epidemiology and management of root diseases. Springer, Berlin Heidelberg New York, pp 293–306

Friesland H, Schrödter H (1988) The analysis of weather factors in epidemiology. In: Kranz J, Rotem J (eds) Experimental techniques in plant disease epidemiology. Springer, Berlin Heidelberg New York, pp 115–134

Frinking HD, Linders EGA (1986) A comparison of two pathosystems: downy mildew on *Spinacia oleracea* and on *Chenopodium album*. Neth J Plant Pathol 92:97–106

Gäumann E (1951) Pflanzliche Infektionslehre, 2nd edn. Birkhäuser, Basel

Garreth-Jones D (ed) (1998) The epidemiology of plant diseases. Kluwer, Dordrecht

Garrett KA (1997) Use of statistical tests of equivalence (bioequivalence tests) in plant pathology. Phytopathology 87:372–374

Gassert W (1978) Rinde und mumifizierte Kirschen als Quelle des Primärinokulums der Kaffeekirschen-Krankheit in Äthiopien. Z Pflanzenkr Pflanzenschutz 75:30–40

Geiger HH, Heun M (1989) Genetics of quantitative resistance to fungal diseases. Annu Rev Phytopathol 27:317–341

Georgopoulos SG, Dekker J (1982) Detection and measurement of fungicide-resistance. In: Dekker J, Georgopoulos SG (eds) Fungicide resistance in plant protection. PUDOC, Wageningen, pp 24–31

Gilligan CA (1983) Modeling of soilborne pathogens. Annu Rev Phytopathol 21:45–64

Gilligan CA (1985) Construction of temporal models. III. Disease progress of soil-borne pathogens. In: Gilligan CA (ed) Mathematical modeling of crop disease, vol 3. Advances in plant pathology. Academic Press, London, pp 67–105

Gilligan CA (1988) Analysis of spatial patterns of soilborne pathogens. In: Kranz J, Rotem J (eds) Experimental techniques in plant disease epidemiology. Springer, Berlin Heidelberg New York, pp 85–98

Gilligan CA (1990a) Mathematical modeling and analysis of soilborne pathogens. In: Kranz J (ed) Epidemics of plant diseases, 2nd edn. Springer, Berlin Heidelberg New York, pp 96–142

Gilligan CA (1990b) Comparison of disease progress curves. New Phytol 115:223–242

Gilligan CA (1994) Temporal aspects of the development of root diseases. In: Campbell CL, Benson DM (eds) Epidemiology and management of root diseases. Springer, Berlin Heidelberg New York, pp 148–194

Gilligan CA (2002) An epidemiological framework for disease management. Adv Bot Res (in press)

Gittleman JL, Luh HK (1992) On comparing comparative methods. Annu Rev Ecol Syst 23:383–404

Gottwald TR, Garnsey SM, Borbón J (1998) Increase and patterns of spread of citrus Tristeza virus infections in Costa Rica and Dominican Republic in the presence of the brown citrus aphid, *Toxoptera citricida*. Phytopathology 88:621–636

Gregory PH (1961) The microbiology of the atmosphere, 1st edn. Leonard Hill, London

Gregory PH (1968) Interpreting plant dispersal gradients. Annu Rev Phytopathol 6:189–212

Gregory PH (1973) The microbiology of the atmosphere, 2nd edn. Leonard Hill, London

Gregory LV, Ayers JE, Nelson RR (1981) Reliability of apparent infection rates in epidemiological research. Phytopathol Z 100:135–142

Griffiths E (1981) Iatrogenic plant diseases. Annu Rev Phytopathol 19:69–82

Groth JV, Roelfs AP (1982) The effect of sexual and asexual reproduction on race abundance in cereal rust fungus populations. Phytopathology 72:1503–1507

Habgood RM (1970) Designation of physiological races of plant pathogens. Nature 227:1268–1269

Hamelink J, Hau B, Kranz J (1988) Untersuchungen zur Entwicklung von Befalls-Verlust-Relationen bei Schaderreger-Komplexen an Einzelpflanzen des Maises. Z Pflanzenkr Pflanzenschutz 95:258–269

Hamilton LM, Stakman EC (1967) Time of stem rust appearance on wheat in the Western Mississippi Basin in relation to the development of epidemics. Phytopathology 57:609–614

Hart LP, Fulbright DW, Clayton JL, Ravenscroft AV (1984) Occurrence of *Septoria nodorum* blotch and *S. tritici* in Michigan winter wheat. Plant Dis 68:418–420

Hau B (1985) Epidemiologische Simulatoren als Instrumente der Systemanalyse mit besonderer Berücksichtigung eines Modells des Gerstenmehltaus. Acta Phytomedica 9. Parey, Berlin

Hau B (1988) Anmerkungen zum asymptotischen Verhalten und dem Schwellentheorem bei Modellen für Epidemien von Pflanzenkrankheiten. Z Pflanzenkr Pflanzenschutz 95:372–376

Hau B (1990) Analytical models of plant disease in a changing environment. Annu Rev Phytopathol 28:221–245

Hau B, Kranz J (1977) Ein Vergleich verschiedener Transformationen von Befallskurven. Phytopathol Z 88:53–68

Hau B, Kranz J (1978) Modellrechnung zur Wirkung des Hyperparasiten *Eudarluca caricis* auf Rostepidemien. Z Pflanzenkr Pflanzenschutz 85:131–141

Hau B, Kranz J (1990) Mathematics and statistics for analysis in epidemiology. In: Kranz J (ed) Epidemics of plant diseases, 2nd edn. Springer, Berlin Heidelberg New York, pp 12–52

Hau B, de Vallavieille-Pope C (1998) Wind-dispersed diseases. In: Garreth-Jones D (ed) The epidemiology of plant diseases. Kluwer, Dordrecht, pp 323–347

Hau B, Kranz J, Dengel HJ, Hamelink J (1980) On the development of loss assessment methods in the tropics. In: Teng PS, Krupa SV (eds) Crop loss assessment. Misc Publ Minn Exp Stn, St Paul, Minn

Hau B, Eisensmith S, Kranz J (1985) Construction of temporal models. II. Simulation of aerial epidemics. In: Gilligan CA (ed) Advances in plant pathology, vol 3. Mathematical modelling of crop disease. Academic Press, London, pp 31–65

Hau B, Kranz J, König R (1989) Fehler beim Schätzen von Befallsstärken bei Pflanzenkrankheiten. Z Pflanzenkr Pflanzenschutz 96:649–674

Hau B, Amorim L, Bergamin Filho A (1993) Mathematical functions to describe disease progress curves of double sigmoid pattern. Phytopathology 83:928–932

Hawkins DM, Fatti LP (1984) Exploring multivariate data using the minor principle components. Statistician 33:325–338

Herrman T, Wiese MV (1985) Influence of cultural practices on incidence of foot rot in winter wheat. Plant Dis 69:948–950

Hinde RA (1973) Das Verhalten der Tiere. I. Suhrkamp, Frankfurt/Main

Hirst JM, Stedman DJ (1960) Epidemiology of *Phytophthora infestans*. I. Climate, ecoclimate and phenology of disease outbreak. Ann Appl Biol 48:471–488

Hirst JM, Stedman OJ, Hogg WH (1963) Long-distance spore transport: Method of measurement, vertical spore profiles, and detection of immigrant spores. J Gen Microbiol 48:329–355

Hoffmann GM, Nienhaus F, Poehling HM, Schönbeck F, Weltzien HC, Wilbert H (1984) Lehrbuch der Phytomedizin, 3. Aufl. Blackwell Wissenschaft, Berlin

Horsfall JG (1979) Iatrogenic disease: mechanisms of action. In: Horsfall FG, Cowling EB (eds) Plant disease: an advanced treatise, vol IV. Academic Press, London, pp 343–355

Huber DJ, Watson RD (1974) Nitrogen form and plant disease. Annu Rev Phytopathol 12:139–165

Huber L, Madden LV, Fitt BDL (1998) Rain splash and spore dispersal: a physical prospective. In: Garreth-Jones D (ed) The epidemiology of plant diseases. Kluwer, Dordrecht, pp 348–370

Hughes GL, Madden LV (1993) Using the beta-binomial distribution to describe aggregated patterns of disease incidence. Phytopathology 83:759–763

Hughes G, McRoberts N, Madden LV, Nelson SC (1997) Validating mathematical models for plant disease progress in space and time. IMA J Math Appl Med Biol 14:85–112

Hull R, Adams AN (1968) Groundnut rosette and its assistor virus. Ann. Appl. Biol. 62:139–145

Hunter JE, Schmidt FL (1990) Methods of meta-analysis. Correcting error and bias in research findings. SAGE Publications Inc, Newbury Park CA, London, 592 pp (paperback 1995)

James WC (1974) Assessment of plant diseases and losses. Annu Rev Phytopathol 12:27–48

Jarvis WR (1989) Managing disease in greenhouse crops. Plant Dis 73:190–194

Jeger MJ (1984) Relating disease progress to cumulative numbers of trapped spores: apple powdery mildew and scab epidemics in sprayed and unsprayed orchard plants. Plant Pathol 33:517–523

Jeger MJ (1985) Models in focus expansion. In: MacKenzie DR, Barfield CS, Kennedy GG, Berger RD (eds) The movement and dispersal of agriculturally important biotic agents. Claitor's Publishing Division, Baton Rouge, pp 279–288

Jeger MJ (1986) The potential of analytic compared with simulation approaches in plant disease epidemiology. In: Leonard KJ, Fry WE (eds) Plant disease epidemiology. Population dynamics and management, vol 1. Macmillan, New York, pp 255–281

Jeger MJ (1987) Modelling the dynamics of pathogen populations. In: Wolfe MS, Caten CE (eds) Populations of plant pathogens, their dynamics and genetics. Blackwell, London, pp 91–107

Jeger MJ (ed) (1989) Spatial components of plant disease epidemics. Prentice-Hall, Englewood Cliffs, New Jersey

Jeger MJ (1990) Mathematical modeling of spatial aspects in plant disease epidemics. In: Kranz J (ed) Epidemics of plant diseases, 2nd edn. Springer, Berlin Heidelberg New York, pp 53–95

Jeger MJ, van den Bosch F (1994a) Threshold criteria for model plant disease epidemics. I. Asymptotic results. Phytopathology 84:24–27

Jeger MJ, van den Bosch F (1994b) Threshold criteria for model plant disease epidemics. II. Persistence and endemicity. Phytopathology 84:28–30

Jensen HP, Christensen E, Jörgensen JH (1992) Powdery mildew resistance genes in 127 northwest European spring barley varieties. Plant Breed 108:210–228

Jörg E (1987) Synökologische Untersuchungen über Wechselwirkungen im Agro-Ökosystem Winterweizen. PhD Thesis Univ Giessen

Jörg E, Weissenfels D, Kranz J (1987) Krankheits- und Schädlingsbefall an Haupt- und Nebenhalmen des Weizens. Z Pflanzenkr Pflanzenschutz 94:509–519

Jörg E, Kranz J, Weber A (1990) Synepidemiologische Untersuchungen im Agro-Ökosystem Winterweizen. In: Heitefuss R (ed) Integrierte Pflanzenproduktion. Deutschen Forschungsgemeinschaft. VHC, Weinheim

Johnson R (1984) A critical analysis of durable resistance. Annu Rev Phytopathol 22:309–330

Johnson RA, Wilcoxson RD (1978) Components of slow rusting in barley infected *Puccinia hordei*. Phytopathology 68:1470–1474

Johnson DA, Aldredge JR, Allen JR, Allwine R (1991) Spatial pattern of downy mildew in hop yards during severe and mild epidemics. Phytopathology 81:1369–1374

Johnson KB, Radcliffe EB, Teng PS (1986) Effects of interacting populations of *Alternaria solani*, *Verticillium dahliae* and the potato leafhopper (*Empoasca fabae*). Phytopathology 76:1046–1052

Jones FWG (1980) Some aspects of the epidemiology of plant parasitic nematodes. In: Palti J, Kranz J (eds) Comparative epidemiology. A tool for better disease management. PUDOC, Wageningen, pp 71–92

Journal AG, Huijbregts CJ (1978) Mining geostatistics. Academic Press, London

Kampmeijer P, Zadoks JC (1977) EPIMUL, a simulator of foci and epidemics in mixtures of resistant and susceptible plants, mosaics and multilines. PUDOC, Wageningen

Kato H (1974) Epidemiology of rice blast. Plant Protect Res 7:1–20

Kelber E (1977) Multivariate Schätzmodelle für Ertragsverluste durch *Cercospora beticola* bei Zuckerrüben. Z Pflanzenkr Pflanzenschutz 84:174–186

Khoury W (1989) Vergleichende Untersuchungen zum Einfluss von Wetterfaktoren auf den Befallsverlauf einiger Getreidekrankheiten. PhD Thesis, Univ Giessen

Khoury W, Kranz J (1994) An assessment of Linear Structural Relationships (LISREL) for the analysis of interactions between weather factors in epidemiological studies. Z Pflanzenkr Pflanzenschutz 101:286–296

Kirby RS, Archer WA (1927) Diseases of cereal and forage crops in the United States in 1926. Plant Dis Rep Suppl 53:110–208

Kirste H (1958) Ergebnisse von Krautfäule-Spritzversuchen. Kartoffelbau 9:114–115

Köhler W, Schachtel G, Voleske P (1996) Biostatistik, 2nd edn. Springer, Berlin Heidelberg New York

Koch H (1980) Räumliche Befallsmuster beim echten Mehltau der Gerste. Z Pflanzenkr Pflanzenschutz 87:731–737

Koch H, Hau B (1980) Ein psychologischer Aspekt beim Schätzen von Pflanzenkrankheiten. Z Pflanzenkr Pflanzenschutz 87:533–545

van Kraayennoord CWS, Laundon GF, Spiers AG (1974) Poplar rusts invade New Zealand. Plant Dis Rep 58:423–427

Kranz J (1968a) Eine Analyse von annuellen Epidemien pilzlicher Parasiten. I. Die Befallskurven und ihre Abhängigkeit von einigen Umweltfaktoren. Phytopathol Z 61:59–86

Kranz J (1968b) Eine Analyse von annuellen Epidemien pilzlicher Parasiten. II. Qualitative und quantitative Merkmale der Befallskurven. Phytopathol Z 61:171–190

Kranz J (1968c) Eine Analyse von annuellen Epidemien pilzlicher Parasiten. III. Über Korrelationen zwischen quantitativen Merkmalen von Befallskurven und Ähnlichkeit von Epidemien. Phytopathol Z 61:205–217

Kranz J (1968d) Zur Infektion und Erkrankung der Banane durch *Mycosphaerella musicola* Leach. Z Pflanzenkr Pflanzenschutz 75:518–527

Kranz J (1974a) Comparison of epidemics. Annu Rev Phytopathol 12:355–374

Kranz J (ed) (1974b) Epidemics of plant diseases. Mathematical analysis and modeling. Springer, Berlin Heidelberg New York

Kranz J (1975a) Beziehungen zwischen Blattmasse und Befallsentwicklung. Z Pflanzenkr Pflanzenschutz 82:621–654

Kranz J (1975b) Das Abklingen von Befallskurven. Z Pflanzenkr Pflanzenschutz 82:655–664

Kranz J (1976) Einfluss einiger Blattkrankheiten auf den Abgang der Blätter. Z Pflanzenkr Pflanzenschutz 83:234–237

Kranz J (1977) A study on maximum severity in plant diseases. In: Travaux dédies à G. Viennot-Bourgin. Paris, pp 169–173

Kranz J (1978) Comparative anatomy of epidemics. In: Horsfall JG, Cowling EB (eds) Plant disease, vol II. Academic Press, New York, pp 33–62

Kranz J (1980) Comparative epidemiology: an evaluation of scope, concepts and methods. In: Palti J, Kranz J (eds) Comparative epidemiology. A tool for better disease management. PUDOC, Wageningen, pp 18–28

Kranz J (1987) Raten für die Dynamik von Subpopulationen bei Krankheitserregern. Z Pflanzenkr Pflanzenschutz 94:225–229

Kranz J (1988a) Measuring plant disease. In: Kranz J, Rotem J (eds) Experimental techniques in plant disease epidemiology. Springer, Berlin Heidelberg New York, pp 35–50

Kranz J (1988b) The methodology of comparative epidemiology. In: Kranz J, Rotem J (eds) Experimental techniques in plant disease epidemiology. Springer, Berlin Heidelberg New York, pp 279–289

Kranz J (1990a) Epidemics of plant diseases. Mathematical analysis and modeling, 2nd edn. Springer, Berlin Heidelberg New York

Kranz J (1990b) Fungal diseases in multispecies plant communities. New Phytol 116:383–405

Kranz J (1996) Epidemiologie der Pflanzenkrankheiten. Ulmer, Stuttgart

Kranz J, Aust HJ (1979) Schatten als epidemiologische Einflußgröße beim Gerstenmehltau. Z Pflanzenkr Pflanzenschutz 86:533–545

Kranz J, Hau B (1980) Systems analysis in epidemiology. Annu Rev Phytopathol 18:67–83

Kranz J, Jörg E (1989) The synecological approach in plant disease epidemiology. Rev Trop Plant Pathol 6:27–38

Kranz J, Knapp R (1973) Qualitative und quantitative Unterschiede der Vergesellschaftung parasitischer Pilzarten in verschiedenen Vegetationseinheiten. Phytopathol Z 77:235–251

Kranz J, Lörincz D (1970) Methoden zum automatischen Vergleich epidemischer Abläufe bei Pflanzenkrankheiten. Phytopathol Z 67:225–233

Kranz J, Rotem J (1988) (eds) Experimental techniques in plant disease epidemiology. Springer, Berlin Heidelberg New York

Kuhn TS (1978) Die Entstehung des Neuen. Suhrkamp, Frankfurt/Main

Kushalappa AC, Ludwig A (1982) Calculation of apparent infection rate in plant disease: development of a method to correct for host growth. Phytopathology 72:1373–1377

Lannou C, Mundt CC (1996) Evolution of a pathogen population in host mixtures: rate of emergence of complex races. Theor Appl Genet 94:991–999

Laranjeira FF, Amorim L, Bergamin-Filho A, Berger RD, Hau B (1998) Analise espacial do amarelecimiento fatal do dendereiro como ferramenta para elucidar sua etiologia. Fitopatol Brasil 23:397–403

Larios C, JF, Moreno RAM (1977) Epidemiologia de algunas enfermendades foliares de la yuca en differentes systemas de cultivo. I. Mildiu polvoroso y roña. Turrialbal 26:389–398

Leonard KJ, Thakur RP, Leath S (1988) Incidence of bipolaris and exserohilum species in corn leaves in North Carolina. Plant Dis 72:1034–1038

Limpert E (1987) Spread of barley mildew by wind and its significance for phytopathology and barley cultivation in Europe. In: Boehm G, Leuschner RM (eds) Advances in aerobiology. Proc 3rd Int Conf Aerobiology, Experientia, Basel

Littell RC, Freund RJ, Spector PC (1991) SAS system for linear models. SAS series in statistical applications, SAS Institute, Cary, NC, 329 pp

Lorenz K (1978) Vergleichende Verhaltensforschung. Springer, Berlin Heidelberg New York

Luo Y, Zeng SM (1995) Simulation studies on epidemics of wheat stripe rust (*Puccinia striiformis*) on slow rusting cultivars and analysis of effects of resistance components. Plant Pathol 44:340–349

Madden LV (1986) Statistical analysis and comparison of disease progress curves. In: Leonard KJ, Fry WE (eds) Plant disease epidemiology, vol 1. Population dynamics and management. Macmillan, New York, pp 55–84

Madden LV, Campbell CL (1990) Nonlinear disease progress curves. In: Kranz J (ed) Epidemics of plant diseases. Mathematical analysis and modeling, 2nd edn. Springer, Berlin Heidelberg New York, pp 181–229

Madden LV, Hughes GL (1995a) Some methods allowing for aggregated patterns of disease incidence in the analysis of data from designed experiments. Plant Pathol 44:927–943

Madden LV, Hughes GL (1995b) Plant disease incidence: distributions, heterogeneity, and temporal analysis. Annu Rev Phytopathol 33:529–564

Madden LV, Louie R, Abt JJ, Knoke JK (1982) Evaluation of tests for randomness of infected plants. Phytopathology 72:195–198

Maddison AC, Andebrhan T, Aranzazu F, Silva-Acuna R (1993) Comparative phytosanitation studies. In: Rudgard SA, Maddison AC, Andebrhan T (eds) Disease management in cocoa. Comparative epidemiology of witches broom. Chapman and Hall, London, pp 165–188

Mardia KV, Kent JT, Bibby JM (1979) Multivariate analysis. Academic Press, London

Matheron ME, Pordias M, Matejka JC (1997) Distribution and population dynamics of *Phytophthora citrophthora* and *P. parasitica* in Arizona citrus orchards an effect of fungicides on tree health. Plant Dis 81:1384–1390

McCarthey HA, Fitt BDL (1998) Dispersal of foliar plant pathogens: mechanisms, gradients and spatial patterns. In: Garreth-Jones D (ed) The epidemiology of plant diseases. Kluwer, Dordrecht, pp 138–160

McDermott JM, McDonald BA (1993) Gene flow in plant pathosystems. Annu Rev Phytopathol 31:353–373

McGee DC (1995) Epidemiological approach to disease management through seed technology. Annu Rev Phytopathol 33:449–466

McLean GD, Garrett RG, Ruesink WG (eds) (1986) Plant virus epidemics. Academic Press, Sidney

Mead R, Curnow RJ, Hasted AM (1993) Statistical methods in agriculture and experimental biology, 2nd edn. Chapman and Hall, London

Merchán VM, Kranz J (1986) Wirkung der Blattnässe auf den asexuellen Zyklus des Weizenmehltaus *Erysiphe graminis* f. sp. *tritici* Marchal. Z Pflanzenkr Pflanzenschutz 93:246–254

Merrill W (1967) The oak wilt epidemics in Pennsylvania and West Virginia: an analysis. Phytopathology 57:1206–1210

Milgroom MG, Fry WE (1988) A simulation analysis of the epidemiological principles for fungicide resistance management populations. Phytopathology 78:565–570

Minogue KP (1986) Disease gradients and the spread of disease. In: Leonard KJ, Fry WE (eds) Plant disease epidemiology, vol 1. Population dynamics and management. Macmillan, New York, pp 285–310

Mora-Aguilera G, Nieto-Angel D, Campbell CL, Teliz D (1996) Multivariate comparison of papaya ringspot epidemics. Phytopathology 86:70–78

Morton HV (1994) Chemical management. In: Campbell CL, Benson DM (eds) Epidemiology and management of root diseases. Springer, Berlin Heidelberg New York, pp 276–292

Mouliom-Pefoura A, Lassoudière A, Foko J, Fontem DA (1996) Comparison of development of *Mycosphaerella fijiensis* and *Mycosphaerella musicola* on banana and plantain in the various ecological zones in Cameroon. Plant Dis 80:950–954

Mudita IW, Kushalappa AC (1993) Ineffectiveness of the first fungicide application at different initial disease incidence levels to manage Septoria blight of celery. Plant Dis 77:1081–1084

Nagarajan S, Ajai (1988) Monitoring and mapping long-distance spread of plant pathogens. In: Kranz J, Rotem J (eds) Experimental techniques in plant disease epidemiology. Springer, Berlin Heidelberg New York, pp 243–249

Nagarajan S, Muralidharan K (1995) Dynamics of plant diseases. Allied Publishers Limited, New Delhi

Nagarajan S, Joshi LM, Saari EE (1976) Meteorological conditions associated with long-distance spread of *Puccinia graminis tritici* uredospores in India. Phytopathology 66:198–203

Nagarajan S, Kranz J, Saari EE, Joshi LM (1982) Utility of weather satellite in monitoring cereal rust epidemics. Z Pflanzenkr Pflanzenschutz 89:276–281

Nagarajan S, Seiboldt G, Kranz J, Saari EE, Joshi LM (1984) Monitoring wheat rust epidemics with Landsat-2 satellite. Phytopathology 74:276–281

Navas-Cortés JA, Hau B, Jimenez RM (1998) Effect of sowing date, host cultivar and race of *Fusarium oxysporum* f. sp. *ciceris* on development of *Fusarium wilt* of chickpea. Phytopathology 88:1338–1346

Neher DE, Campbell CL (1992) Underestimating of disease progress rates with the logistic, monomolecular and Gompertz models when maximum disease intensity is less than 100 percent. Phytopathology 82:811–814

Nelson MR, Felix-Gastelum M, Orum TV, Stowell IJ, Myers DE (1994) Geographic information system and geostatistics an the design and validation of virus management programs. Phytopathology 84:898–905

Nelson MR, Orum TV, Jaime-Garcia R, Nadeem A (1999) Applications of geographic information systems and geostatistics in plant disease epidemiology and management. Plant Dis 83:308–319

Nelson SC, Campbell CL (1992) Incidence and patterns of associations of pathogens in a leaf spot complex on white clover in the Piedmont region of North Carolina. Phytopathology 82:1013–1021

Nelson SC, Campbell CL (1993a) Comparative spatial analysis of foliar epidemics on white clover caused by virus, fungi, and a bacterium. Phytopathology 83:288–301

Nelson SC, Campbell CL (1993b) Disease progress, defoliation and spatial patterns in a multiple disease complex on white clover. Phytopathology 83:419–429

Neter J, Wasserman W, Kutner MH (1985) Applied linear statistical models. 2nd edn. Irwin, Homewood, IL

Neyman J (1939) On a new class of contagious distributions applicable in entomology and bacteriology. Ann Math Statist 10:35–37

Ngugi HK, Julian AM, King SB, Peacocke BJ (2000) Epidemiology of sorghum anthracnose (*Colletotrichum sublineolum*) and leaf blight. Plant Pathol 49:129–140

Nicot PC, Rouse DI (1987) Precision and bias of three quantitative soil assays for *Verticillium dahliae*. Phytopathology 77:875–881

Nicot PC, Rouse DI, Yandell BS (1984) Comparison of statistical methods for studying spatial patterns of soilborne plant pathogens in the field. Phytopathology 74:1399–1402

Norton GA (1976) Analysis of decision making in crop protection. Agro-Ecosystem 3:27–44

Notteghem JL (1977) Mesure au champ de la résistance horizontale du riz à *Pyricularia oryzae*. Agron Trop 32:180–195

Ntahimpera N, Wilson LL, Ellis MA, Madden LV (1999) Comparison of rain effects on splash dispersal of three Colletotrichum species infecting strawberry. Phytopathology 89:555–563

Nutter FW Jr, Teng PS, Royer MH (1993a) Terms and concepts for yield, crop loss, and disease thresholds. Plant Dis 77:211–215

Nutter FW Jr, Gleason ML, Jenco JH, Christmas NC (1993b) Assessing the accuracy, inter-rater repeatability, and inter-rater reliability of disease assessment systems. Phytopathology 83:806–812

Oerke E.-C, Dehne HW, Schönbeck F, Weber A (1994) Crop production and crop protection. Elsevier, Amsterdam

Ogawa JM, Hall DH, Koepsel PA (1967) Spread of pathogens within crops as affected by life cycles and environment. In: Gregory PH, Monteith JL (eds) Airborne microbes. Cambridge Univ Press, London, pp 247–266

Ohl L (1991) Die Fitness von Pathotypen des Gerstenmehltaus und deren Verhalten im Konkurrenzmodell. PhD Thesis, Univ Giessen

Onstad DW, Kornkven EA (1992) Persistence and endemicity of pathogens in plant populations over time and space. Phytopathology 82:561–566

Østergard H, Hovmøller MS (1991) Gametic desequilibria between virulence genes in barley powdery mildew populations in relation to selection and recombination. I. Models. Plant Pathol 40:166–177

Østergard H, Shaw MW (1996) Linear models are inappropriate to estimate relative parasitic fitness of pathogens in heterogeneous host populations. Phytopathology 86:561–562

Pak HA (1992) Das Verhalten fungizidresistenter Stämme des Pilzes *Botrytis cinerea* auf der Weinrebe. PhD Thesis, Univ Giessen

Palti J (1971) Biological characteristics, distribution and control of *Leveillula taurica* (L'v.) Arn. Phytopathol Mediter 10:139–153

Palti J (1981) Cultural practices and infectious crop diseases. Springer, Berlin Heidelberg New York

Palti J, Kranz J (eds) (1980) Comparative epidemiology. A tool for better disease management. PUDOC, Wageningen

Park AW, Gubbins S, Gilligan CA (2001) Invasion and persistence of disease in a spatially structured metapopulation. Oikos 94:162–174

Park EW, Lim SM (1985) Empirical estimation of the asymptote of disease progress curves and the use of the Richards generalized rate parameters for describing disease progress. Phytopathology 75:786–791

Parlevliet JE (1997) Durable resistance. In: Hartleb H, Heitefuá R, Hoppe H-H (eds) Resistance of crop plants against fungi. Fischer, Jena, pp 238–253

Pataky JK, Black MC, Beute MK, Wynne JC (1983) Comparative analysis of *Cylindrocladium* black rot resistance in peanut: greenhouse, microplot and field testing procedures. Phytopathology 73:1615–1620

Patten BC (1971) A primer for ecological modeling and simulation with analog and digital computers. In: Patten BC (ed) Systems analysis and simulation in ecology, vol l. Academic Press, New York, pp 4–131

Paysour RE, Fry WE (1983) Interplot interference: a model for planning field experiments with aerially disseminated pathogens. Phytopathology 73:1014–1020

Pedgley DE (1982) Windborne pests and diseases: meteorology of airborne organisms. Ellis Hoorwood, Chichester

Pedgley DE (1986) Long distance transport of spores. In: Leonard KJ, Fry WE (eds) Plant disease epidemiology, vol 1. Population dynamics and management. Macmillan, New York, pp 346–365

Petraitis PS (1998) How can we compare the importance of ecological processes if we never ask, "Compared to what?". In: Resetarits WJ Jr, Bernardo J (eds) Experimental ecology. Issues and perspectives. Oxford Univ Press, New York, pp 183–201

Pfleger TG, Mundt CC (1998) Wheat leaf rust severity as affected by plant density and species proportion in simple communities of wheat and wild oats. Phytopathology 88:708–714

Phipps PM, Deck SH, Walker DR (1997) Weather-based crop and disease advisory for peanuts in Virginia. Plant Dis 81:236–244

Pigeot I (1999) Bioequivalence trials – status and perspectives. Inform Biometr Epidemiol Med Biol 30:96–109

Pinnschmidt HO, Teng PS, Yong L (1994) Methodology for quantifying rice yield effects of blast. In: Ziegler RS, Leong PS, Teng PS (eds) Rice blast disease. IRRI-CABI, Wallingford, pp 381–408

Pinnschmidt HO, Batchelor WD, Teng PS (1995a) Simulation of multiple species pest damage in rice using CERES-rice. Agric Syst 48:193–222

Pinnschmidt HO, Bonman JM, Kranz J (1995b) Lesion development and sporulation of rice blast. Z Pflanzenkr Pflanzenschutz 102:299–306

Pinstrup-Andersen P, de Londrino N, Infante M (1976) A suggested procedure for estimating yield and production losses in crops. PANS 22:359–365

Plaut JL, Berger RD (1981) Infection rates in three pathosystems epidemics initiated with reduced disease severity. Phytopathology 71:917–921

Pons-Kühnemann J (1994) Struktur und Dynamik der Fungizidresistenz in Populationen von *Erysiphe graminis* f. sp. *hordei* bei verschiedenen Applikationsstrategien mit Triadimenol und Ethirimol. PhD Thesis, Univ Giessen

Popper KR (1973) Objektive Erkenntnis. Ein evolutionärer Entwurf. Hoffmann and Campe, Hamburg (transl from Popper KR, 1972, Objective knowledge. Clarendon Press, Oxford)

Populer C (1972) Les épidémies de l'oidium de l'hevea et la phénologie de son hôte dans le monde. INEAC Ser Sci no 115

Populer C (1978) Changes in host susceptibility in time. In: Horsfall JG, Cowling EB (eds) Plant disease, vol II. Academic Press, London, pp 239–362

Putter CAJ (1980) The management of epidemic levels of endemic diseases under tropical subsistence farming conditions. In: Palti J, Kranz J (eds) Comparative epidemiology. A tool for better disease management. PUDOC, Wageningen, pp 93–103

Raccah B, Irwin ME (1988) Techniques for studying aphid-borne virus epidemiology. In: Kranz J, Rotem J (eds) Experimental techniques in plant disease epidemiology. Springer, Berlin Heidelberg New York, pp 209–222

Rangkuty E (1984) Einfluss einiger systemischer Fungizide auf die Alterszusammensetzung der Kolonien des Gerstenmehltaus. Z Pflanzenkr Pflanzenschutz 91:488–496

Rapilly F (1977) Recherche des facteurs de résistance horizontal à la septoriose de blé: *Septoria nodorum*. Résultats obtenues par la simulation. Ann Phytopathol 9:1–19

Rapilly F (1991) L'épidémiologie en Pathologie Végétale: Mycose Aériennes. INRA, Paris

Rapilly F, Fournet J, Skajennikoff M (1970) Etudes sur l'épidémiologie et la biologie de la rouille jaune du blé, *Puccinia striiformis* Westend. Ann Phytopathol 2:5–31

Reuveni R, Rotem J (1973) Epidemics of *Leveillula taurica* on tomatoes and pepper as affected by the conditions of humidity. Phytopathol Z 76:153–157

Reynolds KL, Neher DA (1997) Statistical comparison of epidemics. In: Francl LJ, Neher DS (eds) Exercises in plant disease epidemiology. APS Press, St Paul, Minn, USA

Richter O, Spikermann U, Lenz F (1991) A new model for plant growth. Gartenbauwissenschaft 56:99–106

Robinson RA (1976) Plant pathosystems. Springer, Berlin Heidelberg New York

Roelfs AP (1985) Epidemiology in North America. In: Bushnell WR, Roelfs AP (eds) The cereal rusts, vol 2. Academic Press, New York

Röhmel J (1999) Some comments on a recent FDA draft guidance on bioequivalence as-
    sessment. Inform Biometr Epidemiol Med Biol 30:122–130
Rotem J (1978) Climatic and weather influences on epidemics. In: Horsfall JB, Cowling
    EB (eds) Plant disease. An advanced treatise, vol II. Academic Press, New York,
    pp 317–337
Rotem J (1988) Techniques of controlled-condition experiments. In: Kranz J, Rotem J (eds)
    Experimental techniques in plant disease epidemiology. Springer, Berlin Heidelberg
    New York, pp 19–31
Rotem J (1994) The genus Alternaria. Biology, epidemiology, and pathogenicity. APS
    Press, St Paul, Minn
Rotem J, Palti J (1969) Irrigation and plant disease. Annu Rev Phytopathol 7:267–288
Rotem J, Palti J (1980) Epidemiological factors as related to plant disease control by cul-
    tural practices. In: Palti J, Kranz J (eds) Comparative epidemiology. A tool for better
    disease management. PUDOC, Wageningen, pp 104–116
Rotem J, Cohen Y, Bashi E (1978) Host and environmental influences on sporulation in
    vivo. Annu Rev Phytopathol 16:83–1001
Rotem J, Bashi E, Kranz J (1983) Studies of crop losses in potato blight caused by *Phy-
    tophthora infestans*. Plant Pathol 32:117–122
Rouse DI (1983) Plant growth models and plant disease epidemiology. In: Kommedahl T,
    Williams PH (eds) Challenging problems in plant health. APS Press, St Paul, Minn
Rouse DI, Mackenzie DR, Nelson RR (1981) A relationship between initial inoculum and
    apparent infection rate in a set of disease progress data for powdery mildew on wheat.
    Phytopathol Z 100:143–149
Royer MH, Nelson RR (1981) The effect of host resistance on relative parasitic fitness of
    Helminthosporium maydis race T. Phytopathology 71:351–354
Royle DJ, Shaw MW, Cooke RJ (1986) Patterns of development of *Septoria nodorum* in
    some winter wheat crops in western Europe. Plant Pathol 35:466–476
Rudgard SA, Andebrhan T, Maddison AC, Schmidt RA (1993) Comparative epidemiology
    studies: introduction. In: Rudgard SA, Maddison T, Andebrhan T (eds) Disease man-
    agement in Cocoa. Comparative epidemiology of witches broom. Chapman and Hall,
    London, pp 25–31
Sache I, de Vallavieille-Pope C (1993) Comparison of the wheat brown and yellow rusts for
    monocyclic sporulation and infection processes, and their polycyclic consequences. J
    Phytopathol 138:55–65
Sache I, de Vallavieille-Pope C (1995) Classification of airborne plant pathogens based on
    sporulation and infection characteristics. Can J Bot 73:1186–1195
Sache, I, Zadoks JC (1995) Life-table analysis of Faba bean rust. Eur J Plant Pathol
    101:431–439
SAS Institute Inc. (1988) User's Guide: Statistic Version 6.03. SAS Institute Inc, Cary, NC
Savary S, Zadoks JC (1992) Analysis of crop loss in the multiple pathosystem groundnut-
    rust-late leaf spot. I. Six experiments. Crop Protect 11:99–109
Savary S, Willocquet L, Elazegui FA, Teng PS, Van Du Pham, Zhu Defeng, Tang Qiyi,
    Huang Shiwen, Lin Xiangquing, Singh HM, Srivastava RK (2000a) Rice pest con-
    straints in tropical Asia: characterisation of injury profiles in relation to production
    situations. Plant Dis 84:341–356
Savary S, Willocquet L, Elazegui FA, Castillo NP, Teng PS (2000b) Rice pest constraints
    in Tropical Asia: quantification of yield losses due to rice pests in a range of produc-
    tion situations. Plant Dis 84:357–369
Schlösser I, Kranz J, Bonman JM (1999) Morphological classification of traditional Philip-
    pine rice cultivars in upland nurseries using cluster analysis methods for recommenda-
    tion, breeding and selection purposes. J Agric Crop Sci 184:165–171

Schlösser I, Kranz J, Bonman JM (2000) Characterization of plant type and epidemiological development in the pathosystem "upland rice-rice blast" (*Pyricularia grisea*) by means of multivariate statistical methods. Z Pflanzenkr Pflanzenschutz 107:12–32

Schmidt RA, Rudgard SA, Maddison AC, Andebrhan T (1993) Comparative epidemiology of the witches broom pathosystem. In: Rudgard SA, Maddison AC, Andebrhan T (eds) Disease management in Cocoa. Comparative epidemiology of witches broom. Chapman and Hall, London, pp 131–155

Schmitt CG, Kingsolver CH, Underwood JF (1959) Epidemiology of stem rust of wheat. I. Wheat stem rust development from inoculation foci of different concentration and spatial arrangement. Plant Dis Rep 43:601–606

Schrödter H (1960) Dispersal by air and water – the flight and landing. In: Horsfall JG, Dimond AE (eds) Plant pathology – an advanced treatise, vol III. Academic Press, New York, pp 170–227

Schrödter H (1965) Methodisches zur Bearbeitung phytometeoropathologischer Untersuchungen, dargestellt am Beispiel der Temperaturrelationen. Phytopathol Z 53:154–166

Schrödter H (1987) Wetter und Pflanzenkrankheiten. Springer, Berlin Heidelberg New York

Schrödter H, Ullrich J (1965) Untersuchungen zur Biometeorologie und Epidemiologie von *Phytophthora infestans* (Mont.) de By. auf mathematisch-statistischer Grundlage. Phytopathol Z 54:87–103

Schuld P (1996) Auswirkungen des Befalls durch drei Blattpathogene (*Uromyces appendiculatus*, *Phaeoisariopsis griseola*, *Colletotrichum lindemuthianum*) auf das Wachstum und den Ertrag von Phaseolus-Bohnen. PhD Thesis, Univ Hannover

Schwarzbach E (1979) A high throughput jet trap for collecting mildew spores on living leaves. Phytopathol Z 94:165–171

Seem RC (1988) The measurement and analysis of the effects of crop development on epidemics. In: Kranz J, Rotem J (eds) Experimental techniques in plant disease epidemiology. Springer, Berlin Heidelberg New York, pp 51–68

Seem RC, Haith DA (1986) Systems analysis in epidemiology. In: Leonard KJ, Fry WE (eds) Plant disease epidemiology, vol 1. Population dynamics and management. Macmillan, New York, pp 232–252

Shtienberg D, Fry WE (1990) Field and computer evaluation of spray-scheduled methods for control of early and late blight of potato. Phytopathology 80:772–777

Skylakakis G (1982) The development and use of models describing outbreaks of resistance to fungicides. Crop Prot 1:244–262

Skylakakis G (1983) Theory and strategy of chemical control. Annu Rev Phytopathol 21:117–135

Sneath PH, Sokal RS (1973) Numerical taxonomy. Freeman, San Francisco

Sokal RS, Sneath PH (1963) Principles of numerical taxonomy. Freeman, San Francisco

Stack RW (1980) Disease progression in common root rot of spring wheat and barley. Can J Plant Pathol 2:187–193

Stähle U, Kranz J, Hau B (1984) Untersuchungen zur Konkurrenzfähigkeit eines einfachen und eines komplexen Pathotypen von *Erysiphe graminis* f.sp. *hordei*. Z Pflanzenkr Pflanzenschutz 91:543–548

Staub T (1991) Fungicide resistance: practical experiences with antiresistance strategies and the role of integrated use. Annu Rev Phytopathol 29:421–442

Steiner KG (1973) Der Einfluss der Fungizidbehandlung auf den Verlauf der Kaffeekirschen-Krankheit (*Colletotrichum coffeanum* Noack). Z Pflanzenkr Pflanzenschutz 80:671–681

Subha Rao KV, Berggren GT, Snow JP (1990) Characterization of leaf rust epidemics in Louisiana. Phytopathology 80:402–410

194

References

Sumner DR, Littrell RH (1974) Influence of tillage, planting date, inoculum survival and mixed populations on epidemiology of Southern corn leaf blight. Phytopathology 64:168–173
Sun P, Zeng SM (1995) Modeling the cultivar-race interactions in inter-regional disease spread. Z Pflanzenkr Pflanzenschutz 102:416–421
Sutton JC, Gillespie TJ, James TDW (1988) Electronic monitoring and use of microprocessors in the field. In: Kranz J, Rotem J (eds) Experimental techniques in plant disease epidemiology. Springer, Berlin Heidelberg New York, pp 99–113
Sylvén E (1968) Threshold values in the economics of insect pest control in agriculture. Nat Swed Inst Plant Prot Contrib 14:69–79
Tanne E, Marcus R, Dubitzky E, Raccah B (1996) Analysis of progress and spatial patterns of corky bark in grapes. Phytopathology 74:389–395
Teng SP (1981) Validation of computer models of plant disease epidemics: a review of philosophy and methodology. Z Pflanzenkr Pflanzenschutz 88:49–63
Teng PS (1985) A comparison of simulation approaches to epidemic modeling. Annu Rev Phytopathol 23:351–379
Teng PS (ed) (1987) Crop loss assessment and pest management. APS Press, St Paul, Minn
Thal WM, Campbell CL (1988) Analysis of progress of alfalfa leaf spot epidemics. Phytopathology 78:389–395 patterns of corky bark in grapes. Plant Dis 80:389–395
Thal WM, Campbell CL, Madden LV (1984) Sensitivity of Weibull model parameter estimates to variation in simulated disease progression data. Phytopathology 74:1425–1430
Thresh JM (1974a) Vector relationships and the development of plant viruses. Phytopathology 64:1050–1056
Thresh JM (1974b) Temporal patterns of virus spread. Annu Rev Phytopathol 12:111–128
Thresh JM (1976) Gradients of plant virus diseases. Ann Appl Biol 82:381–406
Thresh JM (1978) The epidemiology of plant virus diseases. In: Scott PR, Bainbridge A (eds) Plant disease epidemiology. Blackwell, Oxford, pp 79–91
Thresh JM (1980) An ecological approach to the epidemiology of plant virus diseases. In: Palti J, Kranz J (eds) Comparative epidemiology. A tool for better disease management. PUDOC, Wageningen, pp 57–70
Thresh JM (1981) The role of weeds and wild plants in the epidemiology of plant virus diseases. In: Thresh JM (ed) Pests pathogens and vegetation. Pitman Advanced Publishing Program, Boston, pp 53–70
Thresh JM (1982) Cropping practices and virus spread. Annu Rev Phytopathol 20:193–218
Thresh JM (1983) Progress curves of plant virus disease. In: Coaker TH (ed) Advances in applied biology, vol VIII. Academic Press, New York pp 1–85
Thresh JM (1991) The ecology of tropical plant viruses. Plant Pathol 40:324–339
Thresh JM, Owusu GLK, Ollenu LAA (1988) Cocoa swollen shoot: an archetypical crowd disease. Z Pflanzenkr Pflanzenschutz 95:428–446
Tischner H, Bauer G (2000) Monitoring für Getreidekrankheiten in Bayern – bewährte Hilfe zum gezielten Fungizideinsatz. Gesunde Pflanzen 52:254–260
Ullrich J, Schrödter H (1966) Das Problem der Vorhersage des Auftretens der Kartoffelkrautfäule (*Phytophthora infestans*) und die Möglichkeit einer Lösung durch die Negativprognose. Nachrichtenbl Dtsch Pflanzenschutz 18:33–40
Van Bruggen AHC (1995) Plant disease severity in high-input compared to reduced-input and organic farming systems. Plant Dis 79:976–984
Van de Lande HL, Zadoks JC (1999) Spatial patterns of spear rot in oil palm plantations in Surinam. Plant Pathol 48:189–201
Van den Bosch F, Zadoks JC, Metz AJ (1988a) Focus expansion in plant disease. I. The constant rate of focus expansion. Phytopathology 78:54–58
Van den Bosch F, Zadoks JC, Metz AJ (1988b) Focus expansion in plant disease. II. Realistic parameter-sparse models expansion. Phytopathology 78:59–64

Van den Bosch F, Frinking HD, Metz AJ, Zadoks JC (1988c) Focus expansion in plant disease. III. Two experimental examples. Phytopathology 78:919–925

Van der Plank JE (1948) The relation between the size of fields and the spread of plant-disease into them, part I. Crowd diseases. Emp J Exp Agric 16:134–142

Van der Plank JE (1963) Plant diseases: epidemics and control. Academic Press, New York

Van der Plank JE (1968) Disease resistance in plants. Academic Press, New York

Van der Plank JE (1975) Principles in plant infection. Academic Press, New York

Vanderplank JE (1982) Host-pathogen interactions in plant diseases. Academic Press, New York

Vanderplank JE (1984) Disease resistance in plants, 2nd edn. Academic Press, New York

Vitti AJ, Bergamin Filho A, Amorim L, Fegius NC (1995a) Epidemiologia comperativa entre a ferrugem comum e a helmintosporiose do milho. I. Efeito de variáveis climáticas sonre os parâmetricos monociclicos. Summa Phytopathol 21:127–130

Vitti AJ, Bergamin Filho A, Amorim L, Fegius NC (1995b) Epidemiologia comperativa entre a ferrugem comum e a helmintosporiose do milho: II. Desenvolvimiento de epidemias sob condicoes naturais de infeccao. Summa Phytopathol 21:131–133

Waggoner PE (1965) Microclimate and plant disease. Annu Rev Phytopathol 3:103–126

Waggoner PE (1974) Simulation of epidemics. In: Kranz J (ed) Epidemics of plant diseases. Mathematical analysis and modeling. Springer, Berlin Heidelberg New York, pp 137–160

Waggoner PE (1978) Computer simulation of epidemics. In: Horsfall JB, Cowling EB (eds) Plant disease. An advanced treatise, vol II. Academic Press, New York, pp 203–222

Waggoner PE (1986) Progress curves of foliar diseases: their interpretation and use. In: Leonard KJ, Fry WE (eds) Plant disease epidemiology, vol 1. Population dynamics and management. Macmillan, New York, pp 3–37

Waggoner PE (1990) Assembling and using models of epidemics. In: Kranz J (ed) Epidemics of plant diseases. Mathematical analysis and modeling, 2nd edn. Springer, Berlin Heidelberg New York, pp 230–260

Waggoner PE, Berger RD (1987) Defoliation, disease, and growth. Phytopathology 77:393–398

Waggoner PE, Horsfall JG (1969) EPIDEM. A simulator of plant disease written for a computer. Conn Agric Exp Stn Bull 698:1–80

Waggoner PE, Rich S (1981) Lesion distribution, multiple infection, and the logistic increase of plant disease. Proc Natl Acad Sci USA 78:3293–3295

Watt KEF (1966) The nature of systems analysis. In: Watt KEF (ed) Systems analysis in ecology. Academic Press, New York, pp 1–14

Weber A, Kranz J (1994a) Untersuchungen über Nebenwirkungen von Fungiziden auf pilzliche Erkrankungen im Agroökosystem Weizen. Z Pflanzenkr Pflanzenschutz 101:604–616

Weber A, Kranz J (1994b) Einfluss unterdosierter Fungizidanwendung auf die Komplexizität des Befalls im Agroökosystem Winterweizen. In: Heitefuá R (ed) Integrierte Pflanzenproduktion. II. Forschungsbericht der Deutschen Forschungsgemeinschaft. VHC, Weinheim

Weber GE (1996) Modelling interaction between *Erysiphe graminis* and *Septoria nodorum* on wheat. Z Pflanzenkr Pflanzenschutz 103:364–376

Wellek S (1994) Statistische Methoden zum Nachweis von Äquivalenzen. Fischer, Stuttgart

Weltzien HC (1972) Geophytopathology. Annu Rev Phytopathol 10:277–298

Weltzien HC (1988) Use of geophytopathological information. In: Kranz J, Rotem J (eds) Experimental techniques in plant disease epidemiology. Springer, Berlin Heidelberg New York, pp 237–242

Welz G (1988) Analysis of virulence in pathogen populations. In: Kranz J, Rotem J (eds) Experimental techniques in plant disease epidemiology. Springer, Berlin Heidelberg New York, pp 165–178

Welz G, Kranz J (1987) Effects of recombination on races of a barley powdery mildew population. Plant Pathol 36:107–113

Welz G, Kranz J (1997) Effects of resistance on the development of disease in crops. In: Hartleb H, Heitefuß R, Hoppe H-H (eds) Resistance of crop plants against fungi. Fischer, Jena

Wenzel JW (1992) Behavioral homology and phylogeny. Annu Rev Ecol Syst 23:361–381

Westlake WJ (1976) Symmetric confidence intervals for bioequivalence trials. Biometrics 32:741–744

Wiese NV (1982) Crop management by comprehensive appraisal of yield determining variables. Annu Rev Phytopathol 20:419–432

Wiese NV, Herrman T, Grube M (1984) Impact of diseases on wheat in Idaho's Kootenai Valley. Plant Dis 68:421–424

Wohlleber B, Kilian M, Kranz J (1993) Die Variabilität der Fungizidsensitivität innerhalb von Rassen des Gerstenmehltaus, *Erysiphe graminis* f.sp. *hordei*. Z Pflanzenkr Pflanzenschutz 100:460–466

Wolfe MS (1986) Dynamics of the pathogen population in relation to fungicide resistance. In: Dekker J, Georgopoulos G (eds) Fungicide resistance in crop protection. PUDOC, Wageningen

Wolfe MS, Finckh MR (1997) Diversity of host resistance within the crop: effects on host, pathogens and disease. In: Hartleb H, Heitefuß R, Hoppe HH (eds) Resistance of crop plants against fungi. Fischer, Jena

Wolfe MS, Schwarzbach E (1978) Patterns of race changes in powdery mildews. Annu Rev Phytopathol 16:159–180

Wolfenbarger DO (1959) Dispersion of small organisms. Incidence of viruses and pollen; dispersion of fungus spores and insects. Lloydia 22:1–106

Xu XM, Butt DJ (1993) PC-based warning systems for use by apple growers. Bull OEPP 23:595–600

Xu XM, Ridout MS (1998) Effects of initial conditions, sporulation rate, and spore dispersal gradient on the spatio-temporal dynamics of plant disease epidemics. Phytopathology 88:1000–1012

Xu XM, Ridout MS (2000) Stochastic simulation of the spread of race-specific and race-nonspecific aerial fungal pathogens in cultivar mixtures. Plant Pathol 49:207–218

Yang XB, Zeng SM (1992) Detecting patterns of wheat stripe rust pandemics in time and space. Phytopathology 82:571–576

Yang XB, Royer MH, Tschanz AT, Tsai BY (1990) Analysis and quantification of soybean rust epidemics from seventy-three sequential planting experiments. Phytopathology 80:1421–1427

Yarwood CE (1973) Some principles of plant pathology. II. Phytopathology 63:1324–1325

Yeh WH, Bonman JM (1986) Assessment of partial resistance to *Pyricularia oryzae* in six rice cultivars. Plant Pathol 35:319–323

Zadoks JC (1961) Yellow rust on wheat. Studies in epidemiology and physiological specialization. Tijdschr Plantenziekten 67:69–256

Zadoks JC (1967) International dispersal in fungi. Neth J Plant Pathol 73:61–80

Zadoks JC (1971) Systems analysis and the dynamics of epidemics. Phytopathology 61:600–610

Zadoks JC (1978) Methodology of epidemiological research. In: Horsfall JB, Cowling EB (eds) Plant disease. An advanced treatise, vol II. Academic Press, New York, pp 64–96

Zadoks JC (1985) On the conceptual basis of crop loss assessment. Annu Rev Phytopathol 23:455–473

Zadoks JC (2000) Foci, small and large: a special class of biological invasion. In: Dieckmann U, Law R, Metz JAJ (eds) The geometry of ecological interaction. Simplifying spatial complexity. Cambridge Univ Press, Cambridge, pp 292–317

Zadoks JC, Schein RD (1979) Epidemiology and plant disease management. Oxford Univ Press, New York

Zadoks JC, Schein RD (1980) Epidemiology and plant-disease management, the known and the needed. In: Palti J, Kranz J (eds) Comparative epidemiology. A tool for better disease management. PUDOC, Wageningen, pp 1–17

Zadoks JC, Chang TT, Konzak CF (1974) A decimal code for the growth stages of cereals. Eucarpia Bull 7:1–10

Ziegler S, Victor N (1999) Gefahren der Standardmethoden für Meta-Analysen bei Vorliegen von Heterogenität. Inform Biometr Epidemiol Med Biol 30:131–140

SUBJECT INDEX

Druck: Strauss Offsetdruck, Mörlenbach
Verarbeitung: Schäffer, Grünstadt